I0126452

PERFUMES *and* COSMETICS

THEIR

PREPARATION AND MANUFACTURE

A COMPLETE AND PRACTICAL TREATISE

FOR THE USE OF THE PERFUMER AND COSMETIC MANUFACTURER.
COVERING THE ORIGIN AND SELECTION OF ESSENTIAL
OILS AND OTHER PERFUME MATERIALS, THE
COMPOUNDING OF PERFUMES AND THE
PERFUMING OF COSMETICS, ETC.

BY

GEORGE WILLIAM ASKINSON, Dr.Chem.

MANUFACTURING PERFUMER

REVISED WITH IMPORTANT ADDITIONS BY A
CORPS OF EXPERT PRACTICAL PERFUMERS

FULLY ILLUSTRATED

Copyright © 2013 Read Books Ltd.
This book is copyright and may not be
reproduced or copied in any way without
the express permission of the publisher in writing

British Library Cataloguing-in-Publication Data
A catalogue record for this book is available from the
British Library

Essential Oils

Essential oils are also known as volatile oils, ethereal oils, aetherolea, or simply as the 'oil of' the plant from which they are extracted, such as the oil of clove. An oil is 'essential' in the sense that it contains the characteristic fragrance of the plant that it is taken from. Essential oils do not form a distinctive category for any medicinal, pharmacological, or culinary purpose - and they are not essential for health, although they have been used medicinally in history. Although some are suspicious or dismissive towards the use of essential oils in healthcare or pharmacology, essential oils retain considerable popular use, partly in fringe medicine and partly in popular remedies. Therefore it is difficult to obtain reliable references concerning their pharmacological merits.

Medicinal applications proposed by those who sell or use medical oils range from skin treatments to remedies from cancer - and are generally based on historical efficacy. Having said this, some essential oils such as those of juniper and agathosma are valued for their diuretic effects. Other oils, such as clove oil or eugenol were popular for many hundreds of years in dentistry and as antiseptics and local anaesthetics. However as the use of essential oils has declined in evidence based medicine, older text-books are frequently our only sources for information!   Modern works are less inclined to generalise; rather than referring to 'essential oils' as a class at all, they prefer to discuss specific compounds, such as methyl salicylate, rather than 'oil of wintergreen.'

Nevertheless, interest in essential oils has considerably revived in recent decades, with the popularity of aromatherapy, alternative health stores and massage. Generally, the oils are volatized or diluted with a carrier oil to be used in massage, or diffused in the air by a nebulizer, heated over a candle flame, or burned as incense.   Their usage goes way back, and the earliest recorded mention of such methods used to produce essential oils was made by Ibn al-Baitar (1188-1248), an Andalusian physician, pharmacist and chemist. Different oils were claimed to have differing properties; some to have an uplifting and energizing effect on the mind such as grapefruit and jasmine, whilst others such as rose lavender have a reputation as de-stressing and relaxing - and also, usefully, as an insect repellent.

The oils themselves are usually extracted by 'distillation', often by using steam -but some other processes include 'expression' or 'solvent extraction'. Distillation involves raw plant material (be that flowers, leaves, wood, bark, roots, seeds or peel) put into an alembic (distillation apparatus) over water. As the water is heated, the steam passes through the plant material, vaporizing the volatile compounds. The vapours flow through a coil, where they condense back to liquid, which is then collected in the receiving vessel. 'Expression' differs in that it usually merely uses a mechanical or cold press to extract the oil. Most citrus peel oils are made in this way, and due to the relatively large quantities of oil in citrus peel and low cost to grow and harvest the raw materials, citrus-fruit oils are cheaper than most other essential oils. 'Solvent extraction' is perhaps the most difficult of the three methods, and is generally used for flowers, which contain too little volatile oil to undergo expression. Instead, a solvent such as hexane or supercritical carbon dioxide is used to extract the oils.

These techniques have allowed essential oils to be used in all manner of products; from perfumes to cosmetics, soaps - and as flavourings for food and drinks as well as adding scent to incense and household cleaning products. The science, history and folkloric tradition of essential oils is incredibly fascinating - and a still much debated area. We hope the reader is inspired by this book to find out more.

PREFACE

The perfume industry of today is more important than at any time in its long and honored career. Beginning centuries ago, probably in the first dawn of a real civilization, with the preparation of a few odorous gums and woods and their adaptation to the purposes of the toilet, it has gradually expanded until we find it now an industry which touches upon many fields of human endeavor. The preparation of perfumes, using the word in the sense of products used only as sources of delightful odors, is now supplemented to an extraordinary degree by the manufacture of perfumed substances; that is, products which owe their chief value to other properties but which are perfumed in order that their appeal to the user may be heightened. In this connection might be mentioned such diverse yet related articles as soaps, face powders, cold creams, tooth pastes, talcums, etc. In fact, it may be stated that every toilet requisite which lends itself to the process is now perfumed. Closely related to these, at least as regards the means employed, are the many flavoring extracts and innumerable beverages which owe their distinctive appeals to the sense of taste to the use of products which are in many cases identical with those used in the compounding of perfumes. Unfortunately the scope of this book does not permit inclusion of a consideration of this latter class of preparations notwithstanding their close relation to the subject in hand.

One hears many references to the important perfumery business and to the science of perfumes, but all too few to the art of perfuming. For it must never be forgotten that in addition to being a business which calls for the usual amount of business acumen and a science of no little complexity, it differs from other businesses and sciences in being as well an art. Perfumes

cannot be compounded in accordance with definite rules; there
are no such rules. When it comes to the origination of a new
and wonderful odor, the scientist must give way to the artist
whose training and highly developed intuition enable him to
choose his materials and blend them with the skill that produces
an odor which is the delight of millions.

The perfumer has the advantage and disadvantage of appealing to the most sensitive of the human faculties, that of smell;
an advantage because it makes possible a ready appreciation of
the finer qualities of his compositions and a disadvantage because
it compels him to come far closer to ultimate perfection than is
required of those appealing to the less critical senses. To be successful, therefore, he must not only possess a wide special knowledge of the materials he uses, their odors, their properties, their
odor values when used in connection with each other, but he
must know as well how they will be modified by the varied inodorous materials in connection with which they may be used,
such as soaps, powders, etc. In addition to this he must possess
the flair for odor values which makes him a true artist.

Lest the reader become dismayed by these requirements, let it
be said that to few is it given to possess all these advantages. It
is the function of this book to make available to all the special
knowledge of the subject which would otherwise remain the
property of the few, and to offer to those interested hundreds of
perfume formulas which have been compounded by expert perfumers and which have stood the rigorous tests of actual use.

That the user of these formulas may work more intelligently,
the book affords him accurate information regarding the source
and character of his materials and enables him to choose them
intelligently. The aim which has been before the author and
publishers has been to place in the hands of the reader all the
data necessary for carrying out reliable processes for the preparation of the most generally approved simple and compound perfume mixtures, and to give him such additional information re-

garding the origin and properties of his various ingredients as will enable him to always secure the best quality and protect himself against adulterations and fraud.

But the user of perfume materials, whether they be of natural or synthetic origin, must never allow himself to forget that he is dealing with substances which depend for their value to him upon the imponderable quantity of odor value. Chemical tests however accurately carried out are not conclusive, and the nose test must still remain the final one. Since few possess the necessary experience to make this test certain, the best safeguard for the buyer is to always obtain his materials from a house which has the reputation of placing quality above all other considerations. Adulteration is, indeed, common in perfume materials, but there are fortunately some firms whose reputation places them above suspicion.

To one of these the wise buyer always turns, remembering that it is impossible to produce excellent perfumes except from the highest quality materials and that apparent cheapness is usually poor economy since it is almost certain to represent adulteration. The housewife does not buy sugar which is half sand because it is cheaper than the unadulterated, nor can the perfumer afford to purchase "sand" in his materials.

The great extension of the perfume industry earlier referred to must be mainly credited to the great advances made by the organic chemist. Synthetic perfume materials are now available in excellent quality and in wide variety. Many of these duplicate the chief constituents of the essential oils, while others no less valuable to the perfumer have never been found to occur in nature. The result has been to make possible the approximate duplication of the flower oils at a fraction of their cost and to permit the perfumer to work out new odor harmonies by the aid of the new notes provided by the chemist. It is not too much to say that the perfume industry has been revolutionized not only by the wider range of possibilities opened, but by the fact that attractively perfumed products are now within the reach of the average purse, an impossibility if we were still entirely dependent upon the rare and expensive natural oils.

It is still true, however, that the most delightful odors are produced by the almost exclusive use of the best natural products in their composition, and the flower oils will never be supplanted by the products of the laboratory. On the contrary, the two work hand-in-hand and the demand for the natural perfume materials has been heightened by their extensive use in connection with the synthetic perfume materials.

The use of synthetics is, therefore, of the utmost importance to the perfumer, and the subject has been adequately treated in this edition, especially in an added chapter dealing with the benefit to be derived by the employment of the partially finished perfume bases, products which put at the command of the average perfumer the results of a skill and experience far beyond his own attainments.

Another added portion of transcendent importance deals with the most recent advances in the preparation of the natural perfumes. Pomades and concretes which were once the only resource of the perfumer have been largely supplanted by more highly finished products in which the fidelity to the natural flower is improved and which are incomparably more convenient for use.

Still other chapters have been supplied by several expert practical perfumers, who have consented to allow others to benefit by their hard-won experience. The importance of the information contained in these chapters cannot be overestimated.

For the rest, the general plan of the earlier editions has been retained, but the material has been carefully revised to put it in complete accordance with the best modern practice and every effort has been made to make the book of the highest possible value to the American perfumer.

A word of caution may be given as to the formulas. These have been carefully compiled by experts, and in all cases represent perfumes of tested value and quality, but it must be remembered that the results desired can be attained only by strict adherence to the instructions and by the use of high-grade ingredients.   Adulterated materials will not produce the same effects.

Tℍꜰ Aᵤᴛʜᴏʀ.

May, 1922.

CONTENTS

CHAPTER IX

CHAPTER X

CHAPTER XI

CHAPTER XII

CHAPTER XIII

CHAPTER XIV

CHAPTER XV

CHAPTER XVI

CHAPTER XVII

CHAPTER XXXI

CHAPTER XXXII

CHAPTER XXXIII

CHAPTER XXXIV

CHAPTER XXXV

CHAPTER XXXVI

CHAPTER XXXVII

CHAPTER XXXVIII

CHAPTER XXXIX

CHAPTER XL

CHAPTER XLI

CHAPTER XLII

CHAPTER XLIII

CHAPTER XLIV

CHAPTER XLV

CHAPTER XLVI

CHAPTER XLVII

CHAPTER XLVIII

CHAPTER XLIX

CHAPTER L

CHAPTER LI

CHAPTER LII

PERFUMES AND THEIR PREPARATION.

CHAPTER I.

THE HISTORY OF PERFUMERY.

THE gratification of his senses is peculiar to man, and it is to this trait that we are indebted for all the arts. The activities which aimed at the gratification of the eye and ear developed into the creative arts and music, and in like manner human endeavor directed toward the stimulation of the sense of smell has in our time assumed the proportions both of an art and a science; for it was nothing but the advancement of chemistry that made it possible to fix all the pleasant odors offered by nature and to create new perfumes by the artistic combination of these scents. The preparation of perfumes is a very ancient art that is met with among all peoples possessed of any degree of civilization. It is particularly the ancient nations of the Orient which had in truth become masters in the manufacture of numerous perfumes.

The first perfume was the fragrant flower; it has continued to be so to the present day: the sprig of dried lavender flowers which we lay in the clothes-press was probably used for the same purpose by the contemporaries of Aristotle. In the Orient, which we may look upon as the cradle of the art of perfumery, the idea suggested itself early to substitute for the delicious fragrance of the flowers some substances of lasting odor; various sweet-scented resins supplied the ma-

terial for this purpose. The use of these aromatic resins
must have been very extensive: the ancient Egyptians alone
consumed extraordinary quantities for embalming their dead.
How highly the Oriental peoples in general prized perfumes
can be learned from the Bible: the Jews (like the Catholics
to the present day) employed an aromatic gum-resin (olibanum, frankincense) in their religious ceremonies; in the Song
of Solomon mention is made of Indian perfumes, for instance,
cinnamon, spikenard, myrrh, and aloes.

Altogether, incense played a prominent part in the religious ceremonies of the ancient Western Asiatic nations—
among many peoples under a theocratic government it was
even believed to be sinful to use incense for other than religious purposes. The Bible teaches us that Ezekiel and Isaiah
protested against it, and that Moses even prescribed the
preparation of certain kinds of incense for use in the taber-
nacle.

Among the most highly civilized people of antiquity, the
Greeks, a large number of fragrant substances, as well as oils perfumed with them—that is to say, perfumes in the same sense as
we still understand the term—was known; this will be no sur-
prise to those familiar with the culture of this remarkable people.
The odor of violets was the favorite among the Greeks; besides this they used the scent of the different mints, thyme,
marjoram, and other aromatic plants. This was carried so
far as to become a matter of fashion for the Greek fop to use
only certain odors in the form of ointments for the hair, others
for the neck, etc. In order to prevent this luxury which was
carried to such an excess, Solon even promulgated a law that
interdicted the sale of fragrant oils to Athenian men (the law
did not apply to the women).

The Romans, who were the pupils of the Greeks in all the
arts, carried the luxury with perfumes perhaps even farther.
In ancient Rome there was a very numerous guild of per-

fumers called *unguentarii;* they are said to have had a street
to themselves in Capua. A Patrician Roman anointed himself three times daily with precious, sweet-scented oils which
he personally took along into his bath in golden vessels of
exquisite workmanship, so-called nartheciæ. At the funeral
of his wife Poppæa, Nero is said to have used as incense more
odorous substances than could be produced in one year in
Arabia, at that time the only reputed source of perfumes.
This luxury went so far that during the games in the open
amphitheatres the whole air was filled with sweet odors ascending from numerous censers arranged in a circle. The
apartments of well-to-do Romans always contained large and
very valuable urns filled with dried blossoms, to keep the air
permanently perfumed.

Roman extravagance with perfumes was carried to such
an excess that under the consulate of Licinius Crassus a law
was passed which restricted the use of perfumery, there being
good reason to fear that there would not be enough for the
ceremonies in the temples.

With the migration of the almost savage Huns and Goths,
the refinement of morals ceased, progress in civilization was
retarded for centuries, and at the same time the use of perfumes disappeared entirely in Europe; but it was otherwise
in the Orient. As an instance we may mention the prophecy
of Mohammed, who promised in the Koran to the faithful in
paradise the possession of black-eyed houries whose bodies
were composed of the purest musk.

The Arabs, the ancient masters of chemistry, were also the
first founders of the art of perfumery. Thus the Arabian
physician Avicenna, in the tenth century, taught the art of
preparing fragrant waters from leaves, and Sultan Saladin, in
1157, on his triumphal entry, had the walls of the mosque of
Omar washed with rose water.

It was the intercourse with the Orient brought about by

the Crusades that made Europeans again more familiar with the art of perfumery, and a number of new odors rapidly became known.   Italy and France, in those times the representatives of culture, were the countries in which the preparation of perfumes was carried on on a large scale. Thus, for instance, we find the name of a Roman family preserved to the present day because one of its members had combined a sweet-scented powder, called Frangipanni after its inventor, which is still in favor, and because his grandson Mauritius Frangipanni had made the important discovery that by treating this powder with spirit of wine the fragrant substance could be obtained in a fluid form.

The fact has been frequently related and repeated, that Catherine de Medici, the wife of Henry II., had made use of the fashion of perfuming the body for the purpose of ridding herself of objectionable persons, by giving them scented gloves prepared and at the same time poisoned by a Florentine named René (Renato ?). We think this tale to be simply a hair-raising fable—modern chemistry knows no substance the mere touch of which could produce the effect of a fatal poison; and it is scarcely credible that such a material had been known at that time and lost sight of since.

In the sixteenth century, especially at the court of Queen Elizabeth, perfumes were used with great extravagance; in fact, were looked upon as one of the necessaries of life.   This luxury was carried still farther at the courts of the sumptuous kings of France; Louis XV. went so far as to demand every day a different odor for his apartments.   A lady's lover always used the same kind of perfume she did.

It is well known that among the Oriental nations perfumes are used so largely that even food is flavored with rose water, musk, etc.; and Indian and Chinese goods always possess a peculiar aroma which is so characteristic for certain products that it was considered to be a sign of genuineness; this was the case, for instance, with the patchouly odor which always adheres to Indian shawls.

A shawl-maker of Lyons, who had succeeded in perfectly imitating Indian shawls with reference to design and colors, spent a fabulous sum to obtain possession of the plant used by the Indian weavers for perfuming their wares. Despite the great outlay caused by the search for this plant, the manufacturer is said to have done a flourishing business with his " genuine " Indian shawls.

In more recent times the wider range of commerce has resulted in the appearance of many new and valuable natural odorant materials and the scientific study of their extraction and preparation has made them available in more convenient form. The culture of the perfume bearing plants has not been neglected and the yield of perfume has been in many cases notably increased. On the other hand, chemistry has added many new notes to the odor scale and has made the production of cheaper but still desirable perfumes possible. The consequent extension in the use of perfumes has led to placing the industry on a higher and more scientific plane and is in a fair way to make it even more important than it is today.

If empiricism is still the rule in the blending of odorous materials it is only because Science is still incapable of measuring accurately such an imponderable quantity as odor value. Yet it may be predicted that the time is not far off when a rational theory of odor values will be advanced on which can be founded experimental work which will result in placing the art of perfuming on a more scientific basis than it can claim today.

At present France claims supremacy in the perfume field, aided as she is by a plentiful supply of native raw materials and by the natural artistic genius of her people. But the American perfumer yields place to none, and if our tariff policy allows him the unrestricted access to a quality of raw materials such as is enjoyed by his Gallic competitor he can easily hold his own. Even now the demand for imported perfumes is a matter of vogue and not of superior quality.

CHAPTER II.

ABOUT AROMATIC SUBSTANCES IN GENERAL.

WE apply the term perfume—which really means a fumigating material—to those substances which make an agreeable impression upon our sense of smell; the French call them briefly *odeurs*, *i.e.*, odors. The high degree of development at present attained by this industry in France and England is the cause of the fact that all perfumes are generally sold under French or English names, which must be borne in mind by manufacturers in this country.

Perfumes or scents, however, exert not only an agreeable impression on the olfactory organ, but their effect extends to the entire nervous system, which they stimulate; when used in excess, they are apt to cause headache in sensitive persons; the laborers in the chemical factories where these substances are produced on a large scale, occasionally even suffer by reason of their stimulating action on the nerves. For this reason perfumes should never be employed otherwise than in a very dilute condition; this necessity arises from a peculiarity of the odorous substances which when concentrated and pure have by no means a pleasant smell and become fragrant only when highly diluted. Oil of roses, of orange flowers, or of jasmine, in fact nearly all aromatic substances, have an almost disagreeable odor when concentrated; only in an extremely dilute state they yield those delightful scents which we admire so much in the blossoms from which they are derived.

It will be easier to understand the almost incredible productiveness of perfumes if we cite as an instance that a few centigrams of musk placed on a sensitive scale can for years fill a large hall with their characteristic odor without showing an appreciable loss of weight, and still particles must separate from the musk and become evenly diffused through the air of the hall because the odor is perceptible throughout every part of it.

It would be an error, however, were we to assume that all aromatic substances possess the same degree of productiveness; some of them, as for instance the odorous principle of orris root, have a comparatively faint smell—a fact which must be borne in mind in the combination of perfumes. Even odors having a very similar effect on the olfactory nerves differ widely in their intensity; for instance, true oil (attar) of roses possesses an intensity more than twice as great as that of the rose geranium; many authorities agree in giving the proportion as three to eight, the first figure being that of rose oil, the second that of oil of rose geranium. Therefore, in order to produce perfumes of equal intensity (having the same effect on the olfactory nerves), we must dissolve in an equal quantity of the menstruum either three parts by weight of the attar of roses or eight parts of the oil of rose geranium.

In the prescriptions for the preparation of perfumes given in this book, these proportions have been carefully weighed; but it will be the office of the trained olfactory sense of the manufacturer to modify them for the various kinds of perfumery in such a way as to produce a truly harmonious pleasant odor.

Although we know many aromatic substances, we are still in ignorance as to the preparation of certain decidedly agreeable odors. Thus no one at present is able to produce the refreshing odor of the sea borne along on the wind, any more than we are able to reproduce the scent exhaled by the forest, especially after a warm rain; chemistry, though it has done much in the domain of perfumery, has thus far thrown no light upon it. Even certain vegetable odors—for instance, the delightful perfume exhaled by some Aroideæ and Primulaceæ —we cannot as yet preserve unchanged in perfumery. This opens an illimitable field for future activity to the progressive manufacturer.

In a book devoted to the production of perfumes it would certainly be in place to say something about the physiological relations of the olfactory sensations; but unfortunately this interesting part of physiology is still enveloped in great obscurity. All we know positively on this subject is that many particles of the odorous bodies evaporate and must come in contact with the olfactory nerves in order to produce the sensation of odor. There is no lack of experiments seeking to draw a parallel between sensations of smell and those of hearing, and, as is well known, we speak of a harmony and dissonance of odors as we do of tones. Piesse, the renowned perfumer, has even made an attempt to arrange the different odors in a " harmonic scale " having the compass of the piano, and to deduce therefrom a law for the mixture of the several aromatic substances. This attempt, although very ingenious, still lacks a scientific foundation. Piesse endeavors to combine the several scents like tones to produce chords in different scales; the chords of odors are to agree with those of tones. Thus far, however, no proof has been furnished that the olfactory nerve and the acoustic nerve have the same organization, and under this supposition alone could Piesse's system be accepted as correct.

The Division of Aromatic Substances According to their Origin.

The majority of the substances used in perfumery are derived from the vegetable kingdom, but some come from the animal kingdom, and for others which do not occur complete in nature we are indebted to chemistry. As is well

known, most blossoms possess a decided odor, which is ex-·
tremely fragrant in some; yet it is not the blossoms alone,
but in different genera various parts are distinguished by
agreeable odors. In some plants the fragrant substances are
contained in every part, as in different pines and the mints;
in others, only in the fruits (nutmeg, vanilla), while the other
parts are odorless; in certain plants only the rinds of the
fruits contain an aromatic substance (oranges, lemons). In
the Florentine Iris the entire plant is odorless—only its root
stock possesses an agreeable, violet-like scent; while, for instance, in the camphor-tree an aromatic substance exists in
the wood, in the cinnamon laurel in the bark, in the clove-
tree mainly in the closed buds.

But taking the aromatic plants all together, we find that it
is particularly their flowers which contain the finest odors,
and that the majority of perfumes are prepared from their
blossoms.

From the animal kingdom we take for the purposes of perfumery only a very small number of substances, among which,
moreover, some peculiar relation exists; while, for instance,
all men would call the odor of violets, roses, vanilla, etc.,
agreeable, the odor of some animal substances is decidedly
obnoxious to many persons, though others like it—an observation which can be verified often with reference to musk.

With the advancement of science, chemical products find
application in ever increasing numbers; among them are substances which owe their origin directly to the vegetable kingdom, while others, such as nitrobenzol and ionone are only indirectly derived from it.

From what has been stated, we learn that our attention
must be directed particularly to those scents which are derived from the vegetable kingdom. To the manufacturer of
perfumery, however, it is a matter of importance whence the
plants are obtained which he uses for the preparation of the

odors; a very slight change in the soil often makes a great difference in the quality of one and the same species; we see this quite clearly in our ordinary strawberry. While the wild fruit is but small in size, it has a delightful aromatic flavor, and the same species transplanted into gardens attains much greater size but possesses only a faint aroma not to be compared with that of the wild variety. The Lombardian violet is large and beautiful, but the German has a much more pleasant odor. On the other hand, the blossoms of the orange-tree obtained from the plants cultivated in pots cannot be compared with reference to their odor with these growing in the Riviera, the strip of coast land of the Mediterranean from Marseilles to Genoa. Altogether the last-named region and the south of France may be called the true garden of the perfumer; in the neighborhood of Grasse, Cannes, Nice, Monaco, and some other towns, extensive plots of ground are set with aromatic plants such as orange-trees, Acacia farnesiana, jasmine, violets, etc., whose products are elaborated in large, well-appointed chemical factories solely devoted to the extraction of their odors. The proximity of the sea-coast, with its favorable climate almost free from frost, permits the cultivation of southern plants, while in the more elevated parts of the country the adjoining Maritime Alps cause a more changeable climate which adapts them to certain other sweet-scented plants.

The great value of the annual production of the flower farms at Grasse, Cannes and Nice is readily appreciated when it is considered that approximately 3,000,000 kilograms of blooms are picked yearly. Orange flowers, roses, jasmine, violets, tuberoses and acacia flowers are among the most important, with orange flowers easily in the lead. Prices on the flowers and on the perfume oils prepared from them fluctuate too widely to permit any approximation of an average value, but under war and immediate post-war conditions the growers were paid many times the old level of prices, and it is doubtful if the growing demand for flower products will ever permit a return to the old levels.   As a result the flower oils are not to be had in good quality except at high figures.   Even the price of $600 a pound for pure Oil Neroli or orange flower oil which prevailed during 1921 was low in view of the prices paid for the blooms by distillers.

The supremacy of the Grasse region in flower products is unquestioned, but even in more northern regions we find extensive cultivation of aromatic plants which are suited to the climate.   In England, for instance, lavender, crisp mint and peppermint are cultivated on a large scale for their perfume, and English peppermint is famous for its superior quality.   English lavender is equally well known, though in importance it is somewhat overshadowed by that grown in the Maritime Alps.

Italy is chiefly important for lemon, orange, bergamot and mandarin oils, though the latter is of minor importance.   Strangely enough, despite the extensive cultivation of orange-trees in Italy, Italian Neroli is of little importance and has a poor reputation for quality.   Many other oils and odorous gums and resins come from the various semi-tropical regions, and several such as eucalyptus from Australia.   Japan, through its ownership of Formosa, practically controls the camphor supply.

Otto of Rose is one of the flower products in which Grasse can claim no supremacy.   Bulgarian Otto has the highest reputation for quality, and in volume dominates the market.   That from Turkey is regarded as distinctly inferior and subject to frequent and extensive adulteration.

In so far as the United States is concerned, the culture of flowers for perfume production has proved of little importance. Several costly and unfortunate experiments have proved that even in the most favored localities the plan is not feasible owing to the fact that climatic conditions do not permit the flowers to develop the proper amount of essential oils.   Beautiful as the blooms may be, the results of extraction or distillation are disappointing.

We have, however, a truly American essential oil industry which is suited to conditions and which is thriving.   Oils of wintergreen, sweet birch, sassafras, peppermint, and others are produced here economically and are exported in large amounts. The further development of this industry offers far greater possibilities than chimerical attempts to establish a competition to Grasse.

For the majority of our natural perfume materials we will always be dependent on countries which are by nature more suited to the growth of the corresponding plants.

As indicated above, the odors used in perfumery may be divided into three distinct groups according to their origin. These groups are:

1. Odors of vegetable origin.
2. Odors of animal origin.
3. Odors of artificial origin—chemical products.

Before describing the preparation of true perfumes, it is necessary to become acquainted with the several raw materials required in their manufacture; that is to say, the simple odorous substances, their origin, their preparation, and their peculiar qualities. Besides these odorous raw materials, the art of perfumery makes use of a number of chemical and mineral products, whose quality largely influences that of the perfume to be made. These, therefore, likewise call for an appropriate description. Among these auxiliary substances are alcohol, glycerin, fixed oils, and solid fats, which play an important part not only in the preparation of the perfumes, but also enter into the composition of many. The liquid handkerchief perfumes always contain a large quantity of alcohol, the scented hair oils consist largely of fixed oils, while solid fats of animal or vegetable origin occur in the so-called pomades. As we shall see, the actual odors, owing to their extraordinary productiveness, constitute generally only a small percentage of the perfumes; the greatest bulk is usually either alcohol, fixed oil, or solid fat.

CHAPTER III.

ODORS FROM THE VEGETABLE KINGDOM.

The origin and function of the aromatic principles occurring in plants has long been a matter of controversy among plant chemists. The old idea that these substances represented a by-product of the plant and served no useful function, being merely deposited in certain glands, has been thoroughly controverted. Chemical and botanical investigations have clearly established the fact that the essential oils represent but a single phase in the plant metabolism and play their part in carrying on the life processes of the organism. It has been demonstrated that the odorant principles are formed by the oxidation in the plant of inodorous substances and that they are subject to further oxidation which converts them into resinous material or sometimes into plant pigments, their oxidation furnishing energy for the natural changes of the plant or flower.

In some cases it has been shown by Dr. Eugene Charabot, premier perfume chemist of France, that the essential oils are set free at a certain stage of the plant metabolism by the hydrolysis of complex glucosides of which they form a part. This hydrolysis may take place in different sections of different plants, and thus we have plants whose perfume is only in the blossom while in others it is confined to the leaves, while in still others it permeates the whole organism. This also explains why certain plants contain different essential oils in the various parts. The orange-tree, for example, gives Oil Neroli from its blossoms, Oil Orange from the peel of the ripe fruits, and Oil Petitgrain from the leaves and twigs.

The odorous substances occurring in the vegetable kingdom are either essential oils, better called volatile oils, balsams, or resins. Speaking generally, the latter classes are of far less importance in the perfume industry and represent more highly oxidized products than do the volatile oils. They are apparently formed in the plants by the oxidation and polymerization of substances of the volatile oil class, and are accordingly of less delicate odor, of a firmer consistency and more highly colored. Their chemical composition is also much more complex than is that of the volatile oils, which are mixtures of a few, or more commonly many substances of comparatively simple chemical structure. The difference between the classes is therefore essentially a chemical one, and on this basis we are forced to class camphor, obviously a gum in its physical characteristics, with the volatile oils, which it assuredly does not resemble except in its chemical nature.

The name "essential (or volatile) oils" which we use for this class of substances is due to the fact that the volatile vegetable aromatic substances cause a stain on paper similar to that produced by oils and fats; but the stain made by the former disappears spontaneously after some time, while that due to true oils and fats persists. The disappearance of the stain depends on the evaporation of the vegetable aromatic substances—a quality not possessed by fats. Hence the volatile vegetable aromatic substances, in contradistinction from non-volatile fixed or fatty oils, have been designated as essential or volatile or ethereal oils. Inasmuch as the latter terms are the ordinary trade names for these substances, we are compelled to retain them despite their incorrectness. The French name for essential oils is *essences;* "essence de lavande," for instance, is the French name for essential oil of lavender, and not for an alcoholic solution of the oil, as might be inferred from the usually accepted meaning of the English terms " essence of lavender," " essence of peppermint," etc., which mean solutions of these essential oils in alcohol.

As the localities where the raw materials—that is, the aromatic plants—are cultivated on a large scale naturally constitute the places of manufacture of essential oils, we find in southern France and in England the most extensive factories devoted exclusively to the preparation of perfumes. In the countries named, a favorable influence is exerted, too, by their situation near the sea, as well as by their trade with tropical lands from which additional aromatic plants are imported.

We have stated above that the manufacture of essential oils forms almost a monopoly in France and England; but there is no doubt that this country (the United States) likewise has an essential oil industry of no little importance and one which seems destined to grow and flourish.

The Chemical Constitution of Vegetable Aromatic Substances.

The first essential for the perfumer who desires to use his materials intelligently is a knowledge of their origin and of the manner in which they are prepared. Only by the aid of this knowledge will he be in a position to appreciate differences in quality. Such knowledge will likewise be helpful as showing the relationships between the various oils. Scarcely secondary to this comes full information regarding the constituents of the natural perfume materials. This is of especial advantage on account of the fact that many of them are now available as synthetics, which products may sometimes be substituted for or used with the more expensive natural oils. It is therefore desirable to begin with a discussion of the chemistry of the essential oils and their sources and origin.

The sources of the odors derived from the vegetable kingdom can be divided, as stated above, into so-called essential oils, balsams, gum-resins or soft resins, and hard resins. Since the latter bear a certain relation to the essential oils from which they are formed through chemical combinations, we must consider them first.

The flowers, the fruits and their rinds, or even the wood of some plants, form the receptacles of essential oils; if they are liquid they are called essential oils *par excellence;* if they are firm they are called camphors. Besides, there are intermediate states between them: oil of rose is always viscid and solidifies even at temperatures considerably above the freezing-point of water (see under Oil of Rose).

The bodies which are generally called volatile oils are mixtures of various substances which may be divided into two classes: non-oxygenated and oxygenated bodies. The former are hydrocarbons of the class commonly known as terpenes and are of little importance as perfume materials. On the contrary, they act mainly as diluents of the really odorous oxygenated substances. This is well illustrated by lemon and orange oils, which contain an unusually high proportion of terpenes. Terpeneless lemon and orange oils are now prepared by the partial elimination of the hydrocarbons, and are estimated as twenty-five to thirty times as strong in flavor and perfume strength as the natural oils.

The oxygenated bodies are probably formed in the plant by the oxidation of the hydrocarbons as has already been stated. Among them are numbered alcohols, formed by the first step in oxidation, aldehydes and ketones formed by the further oxidation of the alcohols, acids formed from the aldehydes by a further addition of oxygen, esters derived by a combination of alcohols and acids, and ethers which are derived from alcohols by the elimination of water.

With these and still other more complex chemical reactions not only possible but actually taking place it is not surprising that the plant perfumes represent mixtures of a wide variety of different chemical substances, and are correspondingly hard to duplicate since many of the important constituents are present in only minute amounts.

According to their physical qualities, essential oils may be described as fluids of a specific narcotic odor, colorless but very refractive, and easily inflammable. Only a few essential oils can be produced in such a state of purity as to appear perfectly colorless; usually they are more or less dark yellow in color, and some even possess a characteristic tint; thus oil of acacia is reddish-brown, oils of rose and absinth are green, oil of chamomile is blue. But a simple experiment will show that the color is not inseparably connected with the oil, for certain tinted oils can be obtained perfectly colorless by being distilled with another less volatile oil which retains the coloring matter.

The boiling-point of essential oils is in general very high

—between 160° and 288° of the centigrade thermometer (C.), or 320° to 550° F. The fact that we smell the essential oils in aromatic plants so distinctly despite their high boiling-point is an evidence of their exceedingly strong influence on the olfactory nerves.

A peculiar property of essential oils, which is of great importance in their preparation, is that of distilling over in large quantities with steam—both ordinary and superheated—that is, at temperatures at most only slightly exceeding 100° C. or 212° F. For this reason essential oils are usually obtained in this way, since they are but slightly soluble in water. Still, most of the oils dissolve in water in sufficient amount to impart to it their characteristic odor and thus to render it often very fragrant. Aqua Naphæ triplex (orange-flower water), rose water, etc., are such as have been distilled over with the essential oils, contain a small quantity of the latter in solution, and hence have a very agreeable odor.

All essential oils dissolve readily in strong alcohol, petroleum ether, benzol, bisulphide of carbon, in liquid and solid fats, in glycerin, etc.; we shall again recur to this important subject under the head of the preparation of the essential oils.

If a freshly prepared essential oil is at once excluded from the air by being placed in hermetically sealed vessels which it completely fills, and is kept from the light, the oil will remain unchanged for any length of time. But if an essential oil is exposed to the air, a peculiar, chemical alteration begins, which proceeds more rapidly and obviously if direct light acts upon the oil at the same time. The odor becomes less intense, the oil grows darker in color and more viscous, and also acquires a peculiar quality: it has a strong bleaching effect which is easily seen on the cork closing the bottle, which is beautifully bleached. After a certain time the oil changes to a viscid, less odorous mass, into balsam, and the latter, after

2

the prolonged influence of the air, finally changes into a brownish, odorless substance, into resin.

These remarkable physical and chemical alterations depend on the fact that the essential oil absorbs oxygen from the air, which it puts into a peculiar condition in which it exerts increased chemical activity and is termed ozonized oxygen. One of the most marked of these effects is the uncommonly strong bleaching power of ozonized or active oxygen. When an essential oil that has altered so far as to contain ozonized oxygen—which is shown by its bleaching vegetable coloring matters such as the juice of cherries, red beets, tincture of litmus, etc., agitated with it—is cooled, we notice the separation from it of a usually crystalline, colorless, and odorless body called stearopten, while the remaining liquid part is called elæopten. Stearopten always contains oxygen, while elæopten still consists only of carbon and hydrogen.

In the formation of the stearopten we distinctly see the beginning process of resinification, which, therefore, is nothing but an oxidation (combination of the essential oil with oxygen). It should, however, be stated that as to many essential oils this is not proven by actual observation. Many of them are not known to us as naturally existing without any stearopten. Balsams are essential oils which have to a great extent changed into resin, which they contain in solution, and thereby have become more or less viscid. If the process of oxidation goes still farther, eventually the greater portion of the essential oil becomes oxidized, the entire mass grows firm, and then possesses only a very faint odor which is due to the last remnants of the unchanged essential oil.

Since aromatic substances during evaporation become mixed with air, it appears probable that they act upon the olfactory nerves only at the moment when they become oxidized.

The entire process of resinification of oil of turpentine can be followed very clearly on the pitch pine (Pinus austriaca, or other species of Pinus), just as oil of turpentine in general can be taken as an example of an essential oil on which the peculiarities of the non-oxygenated essential oils may be easily studied.   In many localities the pitch pine is partly deprived of its bark when it has reached a certain age.   From the trunk exudes oil of turpentine which in the air becomes more and more viscid by the absorption of oxygen and changes into balsam, called turpentine.   The latter is collected and distilled with water, when the unchanged oil of turpentine passes over with the steam, while the odorless resin (rosin or colophony) remains behind in the stills.

The above-mentioned qualities of the essential oils indicate naturally how those used in perfumery, which are often very costly, are to be preserved.   For this purpose small strong bottles should be chosen which are closed with well-fitting glass stoppers, over which is applied a glass capsule ground to fit tightly over the neck of the bottle.   *These bottles should always be completely filled* (hence small bottles should be selected), *and kept tightly closed, in the dark.*   As the action of oxygen is retarded by low temperatures, it is advisable to keep bottles containing essential oils in a cool cellar.   But care must be had never to pour out an essential oil in the cellar near an open candle light.   The vapors are very apt to take fire, as they are quite inflammable.

As there are a great many aromatic vegetable substances, so there are numerous odors, or, to retain the customary though incorrect appellation, numerous essential oils.   All of these, however, cannot be used in the art of perfumery, as some of them do not possess a pleasant odor, as is the case, for instance, with oil of turpentine. (We may state here, however, that very pure oil of turpentine, distilled from certain Coniferæ, has an agreeable, refreshing odor which at present has found application in perfumery under the title of forest perfume or pine-needle essence.)   Besides, there are numerous essential oils which, while possessing a very pleasant odor, still cannot be used in perfumery except for very cheap preparations, though they are employed in much larger quantities in the manufacture of liqueurs. Such oils are: oil of cumin, fennel, juniper, absinth, etc.

As we shall return to this subject in connection with the essential oils which are used in perfumery in general, we will now consider at greater length the aromatic vegetable substances which are employed for the manufacture of fragrant odors.

CHAPTER IV.

THE AROMATIC VEGETABLE SUBSTANCES EMPLOYED IN PERFUMERY.

EVERY fragrant portion of a plant can be used for the preparation of an aromatic substance, and therefore for the manufacture of a perfume. Hence we are unable, in the following enumeration of the aromatic vegetable substances, to make any claim to absolute completeness; for every new scientific expedition may acquaint us with hitherto unknown plants from which the finest odors may be obtained. We have said above that we have not yet even fixed in our perfumes all the odors of the known aromatic plants, and therefore there is still a large field open to the progressive manufacturer.

In the following pages we must restrict ourselves to the description of those aromatic vegetable substances which are used in the laboratories of the most advanced and scientific perfumers for the manufacture of odors. At the same time we lay particular stress on the fact that the knowledge of these raw materials is a matter of the greatest importance to the manufacturer of perfumes because it enables him to appreciate the differences, often very minute, between fine and inferior ·qualities. Every manufacturer who aims at the production of fine goods must make it the rule to use nothing but the best raw materials.

The price of the latter often appears disproportionately high, since the cheaper grades frequently represent the better qualities more or less generously "stretched" or adulterated to conform to the ideas of unintelligent buyers who look only at price, forgetting that it is impossible to attain truly satisfying results except with pure materials. Not only does the use of inferior qualities of materials militate against the best results, but it makes it impossible to accurately duplicate those once obtained on account of lack of uniformity of materials. It is astonishing what a difference in the delicate odor of the finished perfume may be caused by a minute and almost undetectable amount of impurity in one of the materials used.

In this connection let it be said that the best safeguard for the buyer is to draw his supplies only from a house which has the reputation of never sacrificing quality of materials to any other consideration. This applies to synthetics as well as to natural products.

ALLSPICE.

Latin—Pimenta; *French*—Piment; *German*—Piment; Nelkenpfeffer.

This spice consists of the fruit berries, at first green, later black, of the Eugenia Pimenta, indigenous to Central America and the Antilles. It is chiefly used in the manufacture of liqueurs, less in perfumery, though it may be employed as an addition to certain strong odors, particularly that of oil of bay; it serves very nicely for scenting cheap soap.

ANISE.

Latin—Pimpinella Anisum; *French*—Anis; *German*—Anis.

This well-known plant, which is cultivated in many localities on a large scale, belongs to the Order of Umbelliferæ.

The seeds contain about three per cent of a very aromatic essential oil which finds an application in the manufacture of soap, but is chiefly used as a flavoring for liqueurs.

Good anise must have a light color, an agreeable sweetish odor, and a sharp taste. In order to increase the weight, anise is occasionally moistened with water; such seeds look swollen, are apt to become slimy, and then furnish a less fragrant oil. Anise is not to be confounded with star-anise, which will be mentioned hereafter.

BALM.

Latin—Melissa officinalis; *French*—Melisse; *German*—
Melissenkraut.

Melissa officinalis, an herbaceous plant with large, beautiful flowers, which grows wild in our woods, contains a very sweet-smelling oil in small quantities. This can be extracted by distillation from the fresh herb, and furnishes very fine perfumes.

Oil of Melissa of the market is, however, usually an East Indian oil, derived from Andropogon citratus. See under Citronella.

BAY (SWEET BAY).

Latin—Laurus nobilis; *French*—Laurier; *German*—Lorbeerfrüchte.

The fruits of the bay-tree contain much essential oil which is used less in the manufacture of perfumery than for scenting soap. Venice is the most important point of export. See the next article.

BAY (WEST INDIAN).

Latin—Myrcia acris; *French* – (Huile de) Bay; *German*—
Bay (-Oel).

The essential oil obtained from the leaves of this tree, a native of the West Indies, possesses a very aromatic, refreshing odor somewhat resembling that of allspice.   It is known in the market as bay oil or oil of bay.   During the last decade or so its use has largely extended, and, while formerly almost unknown on the continent of Europe, has become an important article for the perfumer.   An alcoholic distillate, prepared by distilling the fresh leaves with the crude spirit from which rum is otherwise obtained, is known as bay-rum, and is used as a pleasant and refreshing wash for the skin.   Bay-rum may also be made by dissolving the oil, together with certain other ingredients, in alcohol.

BENZOIN.

Latin—Benzoinum; *French*—Benjoin; *German*—Benzoëharz.

This gum-resin, which possesses a pleasant vanilla-like odor, comes from a tree belonging to the Order of Styracaceæ, the Styrax Benzoin, and probably another species of Styrax, indigenous to tropical Asia, especially Siam and Sumatra.   The collection of benzoin is very similar to that of pine resin; the bark of the tree is cut open, the exuding juice is allowed to harden on the trunk, and is thus brought into commerce.   Benzoin differs according to its origin, the age of the tree, etc., and in commerce a number of sorts (Siam, Penang, Palembang, and Sumatra) are distinguished.   As a rule, benzoin comes in lumps ranging in size to that of a child's head.   They are of a light gray color and inclose white, almond-shaped pieces.   The finest quality, known as Siam benzoin after its source, usually is in small pieces (Siam benzoin in tears) which are translucent, light yellow to brown externally, but milky white on fracture, and have a strong vanilla odor.   Less fine but still very good is Siam benzoin in lumps, consisting of large reddish-brown pieces inclosing white particles.   All other kinds mentioned above come from the island of Sumatra, in lumps the size of a fist.   What was formerly known as Calcutta benzoin formed large friable pieces of a dirty reddish-gray color.   Siam as well as Penang benzoin often contains, besides benzoic acid, also cinnamic acid; it is not known why it is not a regular constituent.   The worst quality is sold as "benzoin sorts," consisting of brownish pieces without white spots; they are often mixed with splinters of wood, bast fibres, and fragments of leaves, and can be used only for cheap perfumes.

Good benzoin, besides the qualities named, must have a sweetish and burning sharp taste, it should be very friable, and when heated in a porcelain capsule should emit vapors (benzoic acid) of an acrid taste and a pronounced aromatic odor; it should dissolve completely in strong alcohol.   In perfumery, benzoin serves for the preparation of many odors, washes, and the manufacture of benzoic acid.   The latter will be further discussed under the head of aromatic substances obtained by means of chemistry.

BERGAMOT.

Latin—Citrus Bergamia; *French*—Bergamote; *German*— Bergamottefrüchte.

The bergamot is the fruit of a tree belonging to the Order of Aurantiaceæ, which is cultivated in Calabria.   The tree is unknown in a wild state.   The golden-yellow or greenish-yellow fruits, resembling a lemon in shape, have a bitter and at the same time acid pulp; the thin rind contains a very fragrant oil which is used largely in the manufacture of fine perfumery and soaps, and is exported chiefly from Messina and Palermo.

BITTER ALMONDS.

Latin—Amygdala amara; *French*—Amandes amères; *German*—Bittere Mandeln.

The well-known fruits of the bitter almond-tree (Amygdalus communis, var. amara).   There are no definite botanical differences between the sweet and the bitter almond-tree. The only distinct difference is the character of the respective fruits. The aromatic substance obtained from bitter almonds is not present fully formed in the fruits, but results from the chemical transformation of the amygdalin they contain; the latter body is absent in sweet almonds.

CAJUPUT LEAVES.

Latin—Folia Cajuputi.

The leaves of Melaleuca Cajuputi, a tree found in the Indian and Malay Archipelago, which have an aromatic odor resembling that of cardamoms. In the Orient the leaves are used as incense and for the extraction of the oil they contain.

CAMPHOR WOOD.

Latin—Lignum Camphoræ; French—Bois de camphre; German—Campherholz.

The wood of the Camphor-tree, native of China and Japan, is exceedingly rich in essential oil, the firm, white, and strong-scented camphor. The latter is usually prepared from the wood at the home of the tree, especially in Formosa and Japan, so that the wood hardly forms an article of commerce and is here enumerated only for completeness' sake. In China and in Japan, however, it is largely used for the manufacture of cloth-chests, trunks and wardrobes, as these are never invaded by insects.

CARAWAY SEED.

Latin—Semen Carvi; French—Carvi; German—Kümmel-samen.

This plant, Carum Carvi, which is largely cultivated in Germany, contains in its seeds from four to seven per cent of essential oil which is extracted by distillation. Genuine caraway seed is brownish-yellow, pointed at both ends, quite glabrous on examination with a lens, and marked with five longitudinal ribs. Caraway is occasionally confounded with cumin seed, from Cuminum Cyminum, which is easily recognized with a lens: the seeds of the latter plant have fourteen longitudinal ribs and are hairy. The use of caraway in perfumery is limited to ordinary goods, but in the manufacture of liqueurs it is largely employed.

CASCARILLA BARK.

Latin—Cortex Cascarillæ; *French*—Cascarille; *German*—Cascarillarinde.

This is the bark of a West Indian tree, Croton Eluteria, belonging to the Order of Euphorbiaceæ, native of the Bahamas. It occurs in commerce in the shape of pieces the length and thickness of a finger; externally it is white and fissured, internally of a brown color and resinous. Good qualities should be free from dust and fractured pieces (sifted cascarilla), of a warm aromatic taste, and a very agreeable odor which becomes more marked on being heated. Another variety of cascarilla derived from South Africa, Cascarilla gratissima, has very fragrant leaves which can be used immediately as incense, just as cascarilla in general is employed in perfumery chiefly for fumigating powders and waters.

CASSIE.

Latin — Acacia farnesiana; *French* — Cassie; *German* — Acacie.

The flowers of Acacia farnesiana (Willd.), one of the true acacias, native of the East Indies, which flourishes farther north than the other varieties, cultivated largely in southern France for the delightful odor which resembles that of violets but is more intense. The flowers are collected and made to yield their odorous principle by one of the methods to be described hereafter. The plant which is generally but falsely called Acacia in this country, viz., Robinia pseudoacacia, like-wise bears very fragrant flowers which undoubtedly can be made to yield a perfume by some one of the usual methods; but so far we know of no perfume into which the odor of Robinia flowers enters.   Moreover, it is not alone the flowers of Acacia farnesiana which may be utilized for the preparation of the cassie perfume; the black currant, Ribes niger, contains in its flowers an odor closely resembling the former; this is .actually used in the preparation of an oil sold under the name of " oil of cassie."   The latter plant flourishes in our northern States and would answer as a substitute for Acacia farnesina, which cannot stand our northern winters.

CEDAR WOOD.

Latin—Lignum Cedri; *French*—Bois de cèdre; *German*—Cedernholz.

The wood met with in commerce is derived from the Vir-ginian juniper tree, Juniperus virginiana, which is used in large quantities for inclosing lead pencils.   The chips, the offal from this manufacture, can be employed with advantage for the extraction of the essential oil contained therein.   Long uniform shavings of this wood are also used for fumigation, and the sawdust for cheap sachet powders.   Cedar wood is reddish-brown, fragrant, very soft, and splits easily.   In the perfumery industry it usually passes under the name of the " cedar of Lebanon," although the wood from the last-men-tioned tree (Cedrus libanotica) has quite a different agreeable odor, is very firm, reddish-brown, and of a very bitter taste—qualities by which it is readily distinguished from the other.

CINNAMON.

Latin—Cinnamomum; *French*—Canelle; *German*—Zimmt-rinde.

Cinnamon consists of the bark of the young twigs of the

cinnamon-tree, Cinnamomum zeylanicum, indigenous to Ceylon. Good cinnamon consists of thin, tubular, rolled pieces of bark which are smooth, light brown (darker on fracture), of a pronounced characteristic odor, and a burning and at the same time sweet taste. The most valuable in commerce is that from Ceylon; the thicker bark is less fine.

Chinese cinnamon or cassia (French, Cassie; German, Zimmt-cassia) consists of the bark of the cassia-tree, an undetermined species of Cinnamomum indigenous to Southern China; this is grayish-brown and has the general properties of true cinnamon, but it as well as the oil extracted from it has a less fine odor than cinnamon or oil of cinnamon. A very fine kind of Cinnamon has for a number of years past appeared on the market under the name of Saigon cinnamon. It is very rich in oil, and is exported from Cochin-China. Besides the true oils of cinnamon and cassia, other essential oils are met with in commerce under the names of oil of cinnamon flowers and oil of cinnamon leaves, but their odor is not so fine as that of the former. The so-called cinnamon flowers are the unripe fruits of various cinnamon laurels, collected after the fall of the blossoms. They form brownish cones the length of the nail of the little finger, and furnish an essential oil whose odor resembles that of cinnamon.

CITRON.

Latin—Fructus Citri; *French*—Citron; *German*—Citronen-früchte.

The fruit of a tree, Citrus medica, indigenous to northern India, but largely cultivated in the countries situated around the Mediterranean and in other countries   It is cultivated both for the pleasant acid juice of the fruit and for their fragrant rinds. Only the latter are of value for our purposes. It occurs in European commerce under the name of Citronat or citron peel. Good commercial citron peel should be in quarters and as fresh as possible, which is shown by its softness, the yellow color, and the strong odor. Old peel looks shrunken and brownish and has but little pleasant odor.

CITRON FLOWERS.

Latin—Flores Citri; *French*—Fleurs de citron; *German*—
Citronenblüthen.

The flowers of the citron-tree (Citrus medica) are white, fragrant, and contain a very aromatic essential oil; but as the oil is always extracted from the fresh flowers, the latter do not form an article of commerce.

CHERRYLAUREL LEAVES.

Latin—Folia Laurocerasi; *French*—Laurier-cérise; *German*—Kirschlorbeerblätter.

The leaves of this tree (Prunus Laurocerasus), which is largely cultivated for officinal purposes, furnish an odorous substance completely identical with that contained in bitter almonds, or, rather, formed in them under certain conditions. As the extraction of the odorous substance from bitter almonds is much cheaper, cherrylaurel is but rarely used.

CITRONELLA.

Latin—Andropogon Nardus; *French*—Citronelle; *German*—
Citronella.

This grass, which, like the oil prepared from it, is called citronella, is a native of northern India, and is largely cultivated in Ceylon, where large quantities are worked for the oil; for this reason the grass itself is seldom met with in commerce. Its odor is somewhat similar to that of the Indian lemon grass, that of verbena, and that of several other aromatic plants, in place of which citronella is frequently employed.

Much confusion exists in much of the current literature

regarding the source and synonymy of the Indian grass oils and allied products. The following list contains the most important ones:

1. *Andropogon citratus* DC.—Lemon Grass. The oil is known as Lemon Grass Oil, Indian Verbena Oil or Indian Melissa Oil, or simply Oil of Verbena or Oil of Melissa.

2. *Andropogon laniger* Desf.—This is the Juncus odoratus or Herba Schoenanthi of older pharmacy. No oil is prepared from this.

3. *Andropogon muricatus* Retz.—Cuscus or Vetiver. Source of Oil of Vetiver.

4. *Andropogon nardus* L.—Citronella. Source of Oil of Citronella.

5. *Andropogon Schoenanthus* L.—Ginger Grass. The oil is known as Oil of Ginger Grass, Oil of Geranium Grass, Oil of Indian Geranium or simply Oil of Geranium, also Oil of Rose Geranium ["Rose" is here a corruption of the Hindostanee name of the plant, viz., Rusa], Oil of Rusa Grass, Oil of Rusa, Oil of Palmarosa.—The two terms "Oil of Geranium" and "Oil of Rose Geranium" should be abandoned for this oil, to avoid confusion with the "Oil of (Rose) Geranium" obtained from Pelargonium. See under "Geranium."

CLOVE.

Latin—Caryophylli; *French*—Clous de girofle; *German*—
Nelkengewürz.

This well-known spice comes from a tree, Caryophyllus aromaticus, native of the Moluccas, and largely cultivated at Zanzibar, Pemba, and elsewhere. It consists of the closed buds. The main essential of good quality is the greatest possible freshness, which may be recognized by the cloves being full, heavy, reddish-brown, and of a fatty aspect, and they must contain so much essential oil (about 18 per cent) that when crushed between the fingers the latter should be stained yellowish-brown. Before buying, this test should always be made, and attention paid to the fact whether the whitish dust is present in the wrinkles about the head. We have found in commerce cloves from which the essential oil had been fraudulently extracted with alcohol and hence were worthless; such cloves may be recognized by the faint odor and taste, but especially by the absence of the whitish dust.

CUCUMBER.

Latin—Cucumis sativus; *French*—Concombre; *German*—
Gurke.

The well-known fruits of this kitchen-garden plant, though not strictly sweet-scented, possess a peculiar refreshing odor which has found application in perfumery. Certain products belonging under this head require the odor of cucumber, and therefore this plant is to be included among the aromatic plants in a wider sense.

CULILABAN BARK.

Latin—Cortex Culilavan; *French*—Écorce culilaban; *German*—Kulilabanrinde.

The bark of Cinnamomum Culilavan Nees, a plant indigenous to the Molucca islands, used to occur in commerce in the shape of long, flat pieces of a yellowish-brown color, with an odor like a mixture of cinnamon, sassafras, and clove oils. It is rarely met with now.

DILL.

Latin—Semen Anethi; *French*—Aneth; *German*—Dillsamen.

This plant, Anethum graveolens, which is indigenous to the Mediterranean region and southern Russia, contains in all its parts, particularly in the seeds, an oil of a peculiar odor, which is used as a perfume for soap, also in cheap perfumery, and especially as a flavoring for liqueurs.

ELDER FLOWERS.

Latin—Flores Sambuci; *French*—Sureau; *German*—Hollunderblüthen.

This bush, Sambucus niger, which grows wild in Europe, bears umbellar flowers which are officinal, but contain besides a pleasant odor which can be extracted from them. The odor of the flowers deteriorates on drying, hence in perfumery only the fresh flowers should be used. The American elder (Sambucus canadensis) could easily be used in place of it.

FENNEL (SEED AND HERB).

Latin—Fœniculum; *French*—Fenouil; *German*—Fenchel.

This plant, Fœniculum vulgare, Order Umbelliferæ, is largely cultivated in Europe. It contains an essential oil in all its parts, but especially in the seeds. The plant is rarely used in perfumery, but more frequently in the manufacture of liqueurs. The herb, dried and comminuted, enters into the composition of some cheap sachets.

FRANGIPANNI (see Plumeria).

GERANIUM.

Latin—Pelargonium roseum; *French*—Géranium; *German*—Geranium.

Geranium, or rose geranium, said to have been originally indigenous in South Africa, contains in its leaves an essential oil the odor of which resembles that of oil of rose but which is easily distinguished by its sharper geranium-like odor. The finest quality of Oil Geranium comes from North Africa, and African or Algerian Oil Geranium sells at a consistently higher price than other varieties. Large quantities of the oil also come from the Bourbon Islands and this is sold under the name of Oil Geranium Bourbon or Reunion. The cultivation of the plant in France is also quite extensive.

Turkish Oil Geranium or Palmarosa Oil should be known

only under the latter name, as it is not a true geranium and is distinctly inferior in quality. Users of Oil Geranium in rose odors should employ only the highest qualities if they wish to obtain satisfactory and economical results.

HEDYOSMUM FLOWERS.

On the Antilles there are a number of bushes belonging to the Genus Hedyosmum, Order Chloranthaceæ, whose flowers possess a magnificent, truly intoxicating odor. Thus far these odors seem to have been accessible only to English perfumers. The perfumes sold under this name by Continental manufacturers are merely combinations of different odors.

HELIOTROPE.

Latin—Heliotropium peruvianum; *French*—Héliotrope; *German*—Heliotropenblüthen.

The flowers of this plant, which flourishes well in all temperate or tropic countries, possess a very pleasant odor, about the preparation of which we shall have more to say hereafter. In Europe only French perfumers have manufactured it; according to the author's experiments, however, its extraction presents no more difficulty than that of any other plant.

A synthetic, chemical product, known as piperonal, related to vanillin and cumarin, possesses the odor of the heliotrope in a most remarkable degree. It is therefore much used to imitate the latter. In commerce it is known as heliotropin.

HONEYSUCKLE.

Latin—Flores Loniceræ; *French*—Chèvre-feuille; *German* —Geisblattblüthen.

This well-known climbing plant, Lonicera Caprifolium, found in many of our garden bowers, contains an exceedingly fragrant oil in its numerous flowers, but the extraction of this oil has never proved commercially practical, due either to the expense involved or to the alteration of the odor by the process of distillation or extraction.   The so-called Honeysuckles sold for perfumers' use are more or less close imitations of the natural odor, and while only a very few are really satisfactory those few are of great value, as they make possible the duplication of an odor which would otherwise be lacking in the perfumers' repertoire.

HYSSOP.

Latin—Hyssopus officinalis; *French*—Hyssope; *German*—Ysopkraut.

Hyssop possesses a strong odor, a very bitter taste, and is used only for cheap perfumery, but more frequently in the manufacture of liqueurs.

JASMINE.

Latin—Jasminum odoratissimum; *French*—Jasmin; *German*—Jasminblüthen.

True Jasmine—not to be confounded with German jasmine (Philadelphus coronarius, known here as the mock orange, or the Syringa of cultivation) which is likewise employed in perfumery—flourishes particularly in the coast lands of the Mediterranean, where it is cultivated as a dwarf tree.   The odor obtained from the flowers is one of the finest and most expensive in existence, and for this reason it would be well worth trying the cultivation in our Southern States.   At present nearly all the true jasmine perfume (pomade, extract, etc.) comes from France.

LAVENDER.

Latin—Lavandula vera; *French*—Lavande; *German*—Lavendel.

True lavender, Lavandula vera, is one of the longest used of the various aromatic plants, its leaves being exceedingly popular with the ancient Greeks for their delightfully fresh and lasting odor.   The plant flourishes throughout Central Europe, but the production of Oil Lavender centers in the Maritime Alps and in England.   Not only is the wild lavender picked but the cultivation of the plant is yearly becoming more extensive in France, the industry centering around Grasse.   The Oil Lavender from England is perhaps most highly prized, but the French production is larger and of excellent quality.

Aspic or spike lavender is an oil of similar odor but greatly inferior in quality, being distilled not from the Lavandula vera, but, as the name indicates, from Lavandula spica.   Much of this product comes from Spain and is frequently used as an adulterant for true lavender.

LEMON.

Latin—Citrus Limonum; *French*—Limon; *German*—
Limonenfrüchte.

The fruits of the South European lemon-tree, not to be confounded with citrons, resemble the latter in appearance, but they are smaller, have a more acid taste and a thinner rind. The peel contains an essential oil which is very similar in odor to that of the citron.   Hence the oils of lemon, limetta (from Citrus Limetta), and citron are used for the same purposes; but when the three oils are immediately compared, an experienced olfactory organ perceives a marked difference between them.

LEMON GRASS.

Latin—Andropogon citrates; *French*—Schoenanthe; *German*—
Citronengrass.

This grass, which bears a close resemblance to citronella, is largely cultivated, especially in India and Ceylon, for the essential oil it contains.   The odor of the grass is similar to that of verbena, so that its oil is often used as an adulterant or rather as a substitute for the former.   (Compare the article on "Citronella.")

LILAC.

Latin—Flores Syringæ; *French*—Lilas; *German*—Fliederblüthen.

This plant, Syringa vulgaris, a native of Persia but fully acclimated in Europe and in this country, has very fragrant flowers.

A recently discovered liquid principle, now known as terpineol ($C_{10}H_{17}OH$), which exists in many essential oils, and in these, in the portion boiling between 420° and 424° F., possesses the lilac odor in a most pronounced degree, and to its presence in the lilac flowers the peculiar odor of the latter is, no doubt, due.

By itself terpineol is not a satisfactory lilac perfume, as it lacks the flowery character so essential to a flower odor. It is, however, a most valuable base for the cheaper lilac odors. Many specialties are obtainable in the market which represent more perfectly the true lilac perfume, one of the most valuable of which is lilacine. Other more expensive specialties, such as lilafleur, possess the valuable element of floralcy to a much higher degree, being composed of synthetics blended with the more costly floral products. Thus the perfumer who desires the popular lilac odor is enabled to base his composition on one of these scientifically prepared specialties and so obtain results which would otherwise be difficult if not impossible.

What is said of lilac applies with equal force to many other flower odors which are obtainable only in the form of their synthetic replicas. The fact, however, that an oil bears a flower name does not necessarily mean that it is an accurate duplicate.

MACE.

Latin—Macis; *French*—Macis; *German*—Muscatblüthe.

This substance is the dried arillus covering the fruits of Myristica fragrans, the so-called nutmegs. The tree bearing them is indigenous to a group of islands in the Indian Archipelago and is cultivated especially on the Molucca islands. Although mace is in such close relation with nutmeg, yet, strange to say, the aromatic substance differs decidedly from that of the nut. Mace of good quality forms pieces of orange-yellow color; they are fleshy, usually slit open on one side, have a strong odor, tear with difficulty, and are so oily that when crushed they stain the fingers brownish-yellow. Mace is largely used in the preparation of sachets and particularly for scenting soap. In England, soap scented with mace is well liked.

MAGNOLIA.

Latin—Magnolia grandiflora; *French*—Magnolia; *German*
—Magnoliablüthen.

The magnolia (Magnolia grandiflora), indigenous to the warmer parts of South, Central, and North America, bears large white flowers having a delightful odor which can be extracted by means of petroleum ether. In the same way, truly intoxicating perfumes may be obtained from other varieties of magnolia. In our climate these plants flourish only in conservatories, and in their home no steps have yet been taken to utilize these natural treasures in a proper way; hence European manufacturers invariably produce the perfume called magnolia by combination of different odors.

MARJORAM.

Latin—Herba majoranæ; *French*—Marjolaine; *German*—
Majorankraut.

This plant, Origanum Majorana (vulgare), frequently cultivated in kitchen gardens, possesses in all its parts a strong odor due to an essential oil. The latter, which is quite expensive, is but little used, and probably only for culinary purposes.

"Oil of Origanum" in English-speaking countries is intended to mean Oil of Thyme (from Thymus vulgaris), and never means Oil of Marjoram.

MEADOW SWEET.

Latin—Spiræa ulmaria; *French*—Reine des prés; *German*—
Spierstaude.

This plant is frequent in Europe on damp meadows, and contains an aromatic substance closely allied to oil of wintergreen, which occurs also in the Canadian variety.

MINT.

Latin—Mentha; *French*—Menthe; *German*—Minze.

The varieties of mint claiming our attention are the following: *Mentha piperita*, Peppermint (French: Menthe poivrée; German: Pfefferminze).—*Mentha viridis*, Spearmint (French: Menthe verte; German: Grüne Minze).—*Mentha crispa*, Crisp Mint (French: Menthe crépue [or frisée]; German: Krause Minze).

All of the mints have a pleasant odor; besides the plants named above, we may mention Mentha aquatica, whose odor faintly but distinctly recalls that of musk. Like lavender, Mentha crispa and M. piperita are cultivated particularly in England, and the English oils are the most superior. Mentha piperita is also largely cultivated in the United States. Mentha viridis and its oil are almost exclusively confined to this country.   .

MUSK-SEED.

Latin—Semen Abelmoschi; *French*—Grains d'ambrette;
German—Bisamkörner.

The tree, Hibiscus Abelmoschus, indigenous to Africa and India, bears fruit capsules containing reddish-gray seeds with grooved surface, so-called musk-seeds. They have an odor resembling musk, but much weaker, though it becomes more pronounced when the seeds are bruised. Besides this species of Hibiscus, other plants belonging to the same order are aromatic and are also used in perfumery.

MYRRH.

Latin—Myrrha; *French*—Myrrhe; *German*—Myrrhe.

The gum-resin which we call myrrh has long been known in the East, where it was celebrated as one of the finest perfumes, along with spikenard and frankincense. The tree, Balsamodendron Myrrha (or Commiphora Myrrha Engler) is indigenous to the countries bordering the Red Sea to about 22° N. Lat.; the gum exudes partly spontaneously from the trunk. In European commerce myrrh appears in different sorts; that called myrrha electa or myrrha in lacrimis is the most precious; it forms tears of a golden yellow to brown color, traversed by white veins; they have a pleasant smell. That called myrrha naturalis is inferior, but on being heated develops the characteristic aroma. In commerce a product is sometimes offered by the name of myrrh which is nothing but cherry-tree gum scented with genuine myrrh.

MYRTLE LEAVES.

Latin—Myrtus communis; *French*—Myrte; *German*— Myrtenblätter.

The leaves of this Southern European plant diffuse a pleasant odor; the oil to which it is due can be extracted by distillation; yet the perfumes usually called myrtle are not obtained from the plant, but are made by the combination of several aromatic substances. The aromatic water known, especially in France, as "eau d'anges" is obtained by the distillation of myrtle leaves with water.

NARCISSUS.

Latin—Narcissus poeticus; *French*—Narcisse; *German*—
Narcissenblüthen.

The blossoms of this favorite garden plant, which is cultivated on a large scale near Nice, have a pleasant, almost narcotic odor which may be extracted in various ways; though
the greatest part of the so-called narcissus perfumes are made
artificially.

Another species of Narcissus (Narcissus Jonquilla) is frequently cultivated in warm countries for its pleasant scent;
but the perfumes generally found in the market under the
name of Extract, etc., of Jonquil are artificial compounds.

NUTMEG.

Latin—Myristica; *French*—Muscade; *German*—Muscat-
nüsse.

These nuts are almost spherical in shape, the size of a
small walnut, of a grayish-brown color externally, and usually
coated with a faint whitish-gray covering (which is lime). Internally they are reddish-brown, with white marbled spots.
Good fresh nutmegs should be dense, heavy, and so oily that
when pierced with a needle a drop of oil should follow the
withdrawal of the latter. Nuts which are hollow, wormy, and
of a faint odor cannot be used in perfumery. Oil of nutmeg
is used extensively in perfumery, but is rarely employed pure,
more commonly in combination with other strong odors.

OLIBANUM.

Latin—Olibanum; *French*—Encens; *German*—Weihrauch.

This gum-resin, employed even by the ancient civilized
nations of Asia, especially as incense for religious purposes,
comes from East African trees, various species of Boswellia.
Fine olibanum appears in light yellow tears, very transparent

and hard, whose pleasant though faint odor becomes particularly marked when it is thrown on hot coals. In perfumery olibanum is used almost exclusively for pastils, fumigating powders, etc. Pulverulent olibanum constitutes an inferior quality and is often adulterated with pine resin.

OPOPANAX.

Latin—Resina Opopanax.

The root stock of an umbelliferous plant, indigenous in Syria, now recognized at Balsamodendron Kafal, furnishes a yellow milky sap containing an aromatic resin with an odor resembling that of gum ammoniacum. At least the opopanax now obtainable in the market is derived from this source. True opopanax resin, such as used to reach the market formerly, is now unobtainable, and its true source is yet unknown. Opopanax oil is used in perfumery to some extent.

ORANGE FLOWERS.

Latin—Flores Aurantii; *French*—Fleurs d'oranges; *German*—
Orangenblüthen.

The flowers of the bitter orange tree (Citrus vulgaris), as well as those of the sweet (Citrus Aurantium), contain very fragrant essential oils, which differ in flavor and value according to their source and mode of preparation. See below, under Oil of Orange. The leaves, too, contain a peculiar oil used in perfumery.

ORANGE PEEL.

Latin—Cortex Aurantii; *French*—Ecorce d'oranges; *German*
—Orangenschalen.

The very oily rinds of the orange occur in commerce in a dried form; such peels, however, can be used only in the manufacture of liqueurs; in perfumery nothing but the oil from the fresh rinds is employed, and this is generally obtained by pressure.

It is entirely impractical for the perfumer to attempt the preparation of Oil Orange. There are two varieties, Oil Orange, Bitter, and Oil Orange, Sweet, the better qualities of both being obtained by expression from the fresh peel, while an inferior quality is obtained by distillation. Italian orange oils are regarded as the most desirable, but those from the West Indies are gaining in popularity. California orange oils are not important, as labor costs preclude the possibility of preparing the high grade hand-pressed oils which are obtainable from Italy.

ORIGANUM.

See Marjoram, and Thyme.

ORRIS ROOT.

Latin—Radix Iridis florentinæ; *French*—Iris; *German*—Veil-chenwurzel.

The Florentine sword-lily, Iris florentina, which often grows wild in Italy but is largely cultivated, has a creeping root-stock covered with a brown bark which, however, is peeled from the fresh root. Orris root occurs in commerce in whitish pieces which are sometimes forked; the surface is knotty, and the size may reach the thickness of a thumb and the length of a finger. When fresh, the roots have a disagreeable sharp odor, but on drying they attain an odor which may be said to resemble that of the violet; but on comparing the two odors immediately, a considerable difference is perceptible even to the untrained olfactory sense. Orris root should be as fresh as possible; this may be recognized by its toughness, the great weight, and the white, not yellow color on fracture. It is very frequently used for sachets and for fixing other odors.

PATCHOULY.

Latin—Pogostemon Patchouly; *French*—Patchouly; *German* —Patschulikraut.

Patchouly is an herb indigenous to the East Indies and to part of China; Singapore, Straits Settlements, is one of the centers for the collection and marketing of the herb, and to some extent for the distillation of the oil. The highest grade oil, however, is obtainable from London. Oil Patchouly is used in only small amounts by the perfumer, to whom it is nevertheless a valuable ingredient in many compositions. Quality is of paramount importance if a desirable effect is to be obtained and the small amount used makes it economical to use the very best obtainable, since the difference in price between the best and the ordinary adds but little to the cost of the finished perfume.

This herb has been known very long in Europe, but at present it is imported in large quantities from India; in commerce it occurs in small bundles consisting of stems and leaves (collected before flowering).

PERU BALSAM.

Latin—Balsamum peruvianum; *French*—Beaume du Perou; *German*—Perubalsam.

This balsam, imported from Central America (San Salvador), is derived from Toluifera Pereiræ; incisions are made in the bark and trunk of the tree, from which the balsam exudes. Peru balsam is of a syrupy consistence, thick and viscid, brownish-red in thin, blackish-brown in thick layers. Its taste is pungent, sharp, and bitter, afterward acrid; its odor is somewhat smoky, but agreeable and balsamic. Peru balsam is often sophisticated with fixed oil; this can be readily detected by agitation with alcohol, by which the oil is separated. But if castor oil is the adulterant, this test is not applicable, as castor oil dissolves with equal facility in alcohol.

PINE-APPLE.

Latin—Bromelia Ananas; *French*—Ananas; *German*—Ananas.

The fruits of this plant, originally derived from the East Indies, have a well-known narcotic odor which can be extracted from them.

In commerce we often meet with a chemical product called pine-apple ether which will be described at greater length under the head of chemical products used in perfumery. Pine-apple ether has an odor usually considered to be like that of the fruit, but when the two substances are immediately compared a great difference will be detected. Pine-apple ether finds quite extensive application in confectionery for the preparation of lemonades, punch, ices, etc. If the true pine-apple odor is to be prepared from the fruits, care must be had to use ripe fruits; the unripe or overripe fruits possess a less delicate aroma.

PINK.

Latin—Dianthus Caryophyllus; *French*—Œillet; *German*—
Nelkenblüthen.

Natural Oil Carnation is unfortunately not particularly stable in composition, and it was not until the discovery of the valuable carnation base, Œillet, by Philip Chuit, the Swiss perfume chemist, that a satisfactory carnation type was available to the perfumer. This product and less satisfactory imitations under the same name are now used extensively.

PLUMERIA.

Latin—Plumeria; *French*—Plumeria; *German*—Plumeria-
blüthen.

All the Plumerias, indigenous to the Antilles, contain very fragrant odors in their flowers. To the best of our knowledge, these odors have not yet been extracted from the flowers, and all the perfumes sold under this name (sometimes also called Frangipanni) are merely combinations of different odors

RESEDA (MIGNONETTE).

Latin—Reseda odorata; *French* — Mignonette; *German*—
Reseda.

This herbaceous plant, probably indigenous to northern
Africa, but long domesticated in Europe and cultivated in
gardens, is well known for its refreshing odor. The latter,
however, is very difficult to extract and is yielded only to the
method of absorption (enfleurage). The true odor of reseda,
owing to the mode of its preparation, is very expensive, and
for this reason nearly all perfumes sold under this name are
produced from other aromatic substances.

RHODIUM.

Latin—Lignum Rhodii; *French*—Bois de rose; *German*—
Rosenholz.

This is derived from two climbing plants, Convolvulus scoparius and Convolvulus floridus, indigenous to the Canary
islands, and is the root wood of these plants. Its odor resembles that of the rose, and the wood is frequently used for cheap
sachets and for the extraction of the contained essential oil
which was formerly (before oil of rose geranium was made on
the large scale) employed for the adulteration of genuine oil
of rose.

ROSE.

Latin—Rosa; *French*—Rose; *German*—Rosenblüthen.

Roses of many types and colors are known, but the best Otto
of Rose is the product of the glorious red roses of Bulgaria,
where the ancient industry of rose culture still centers around
Kazanlik. The Bulgarian Otto of Rose d'Or is recognized as
the standard. Rose culture is also carried on in several parts of
Turkey, but the Turkish Otto is inferior in quality and subject
to adulteration. Grasse is also a center of rose culture and

produces an Otto of Rose which is of excellent quality, far superior to the Turkish and in some cases equal to the best Bulgarian. Attempts have been made to produce this valuable essential oil here, but have uniformly failed owing to the smaller amount of the oil found in American-grown roses.

Owing to the high cost of the natural product, many imitations of Oil of Rose are on the market, some of which are made by "stretching" the natural oil and others by blending the aromatic substances which are known to be present in true rose oils. The delicate Wild Rose odor is made by blending the natural Otto of Rose with other products.

ROSEMARY.

Latin—Rosmarinus officinalis; *French*—Romarin; *German*—Rosmarin.

This plant, indigenous to Southern and Central Europe, contains pretty large quantities of an aromatic oil in its leaves and flowers; the oil has a refreshing odor and therefore is frequently added in small amounts to fine perfumes.

RUE.

Latin—Ruta graveolens; *French*—Rue; *German*—Raute.

This plant, cultivated in our gardens and also growing wild here, has long been employed for its strong odor; in perfumery rue, in a dry state as well as its oil, is occasionally used.

SAGE.

Latin—Salvia officinalis; *French*—Sauge; *German*—Salbei.

All varieties of sage, the one named being found most frequently growing wild in the meadows of Southern Europe, and extensively cultivated in Europe and in this country, possess a very agreeable, refreshing odor which adheres for a long time even to the dried leaves; these are therefore very suitable for sachets, tooth powders, etc.

SANTAL WOOD.

Latin—Santalum album; *French*—Santal; *German*—Santal-
holz.

The tree from which this wood is derived is indigenous to
Eastern Asia, to the Sunda Islands. The wood is soft, very
fragrant, and is also erroneously called sandal wood. The
latter is of a dark reddish-brown color, not fragrant, and is
derived from Pterocarpus santalinus, a tree indigenous to
Southern India, and the Philippine Islands; it is of value to the
dyer and the cabinet-maker, but to the perfumer only for coloring some tinctures. For the purposes of perfumery use
can be made only of santal wood (white or yellow santal
wood) which possesses a very pleasant odor resembling that
of oil of rose. Formerly essential oil of santal was employed
for the adulteration of oil of rose. White and yellow santal
wood comes from the same tree—the former from the smaller
trunks of Santalum album.

SASSAFRAS

Latin—Lignum Sassafras; *French*—Sassafras; *German*—Sas-
safrasholz.

Sassafras wood, derived from the root of the American
tree Sassafras officinalis, appears in commerce in large bundles. It has a strong peculiar odor; in the bark of the root
the odor is even more marked. In the European drug trade
Sassafras saw dust is also met with, but this is not rarely
mixed with pine saw dust which has been moistened with
fennel water and again dried. In perfumery sassafras wood
is less used for the manufacture of volatile odors than for
scenting soap. Since the principal constituent of oil of sassafras, viz., safrol, has been found to be contained in the crude
oil of Japanese camphor, the latter has to a very large extent
taken the place of the natural oil.

SPIKENARD.

Latin—Nardostachys Jatamansi; *French*—Spic-nard; *German* —Nardenkraut.

This plant, belonging to the Order of Valerianaceæ, which generally possess a strong and more or less unpleasant odor, forms one of the main objects of Oriental perfumery; in the East Indies, where the plant grows wild on the mountains, the odor is held about in the same estimation as that of roses, violets, etc., in Europe. Spikenard was probably known to the ancient Babylonians and Assyrians, for in the Bible, in the Song of Solomon, we find this plant repeatedly mentioned and praised for its pleasant odor. As the odor of spikenard is not appreciated in Europe, the plant is rarely met with in commerce. All parts of the plant are aromatic, but use is chiefly made of the root, consisting of fine fibres which are tied in bundles the thickness of a finger.

STAR-ANISE.

Latin—Illicium; Semen Anisi stellati; *French*—Badiane; *German*—Sternanis.

Star-anise occurs in commerce in the form of eight-chambered capsules, each compartment containing one glossy seed, and is derived from a Chinese tree, Illicium anisatum. The fruits are brown, woody; the seed has a sweetish taste and an odor resembling that of anise. Outside of perfumery star-anise is used in the manufacture of liqueurs. Recently a drug has appeared in commerce under the name of star-anise which possesses poisonous qualities, and is derived from another variety of Illicium (Illicium religiosum). While this may be of no consequence to the perfumer, it is important to the manufacturer of liqueurs who always uses star-anise for fine goods and never oil of anise.

Storax.

Latin—Styrax; *French*—Styrax; *German*—Storax.

This product which belongs among the balsams is derived from a small tree, Liquidambar orientalis, and is obtained from the bark by heating with water, and also by pressure. It forms a viscid mass like turpentine, has a gray color, a burning sharp taste, an agreeable odor, and is easily soluble in strong alcohol; but the odor becomes pleasant only after the solution is highly diluted. Storax has the peculiar property of binding different, very delicate odors, to render them less fugitive, and for this reason finds frequent application in perfumery.

Oriental storax should not be confounded with American storax which occurs in commerce under the name of Sweet Gum, Gum Wax, or Liquidamber, and is derived from Liquidambar styraciflua. It is quite a thick transparent liquid, light yellow, gradually becoming more and more solid and darker colored, but is often used in place of the former, though its odor is less fine.

Sumbul Root.

Latin—Radix Sumbul; *French*—Soumboul; *German*—Moschuswurzel.

The Sumbul plant (Ferula Sumbul), indigenous to Turkestan and adjoining countries, has a light brown root covered with thin fibres, which has a penetrating odor of musk. Owing to this quality it is frequently employed in perfumery, especially for sachets. In commerce a distinction is made between East Indian and Bokharian or Russian sumbul, due to the different routes by which the article arrives. The latter, which possesses the strongest odor, probably because it reaches the market in a fresher state, is the most valuable.

4

SWEET ALMONDS.

Latin—Amygdala dulcis; *French*—Amandes douces; *German*
—Süsse Mandeln.

The almond-tree, Amygdalus communis, occurs in two
varieties, undistinguishable by botanical characteristics. One
bears sweet, the other bitter fruits (comp. Bitter almonds,
page 24). Both are odorless and contain much fixed oil.
The special odor of bitter almonds forms only in consequence
of the decomposition of a peculiar body (amygdalin), present
in bitter almonds, when it comes in contact with water. Good
almonds are full, juicy, light brown, without wrinkles, and
have a sweet mild taste. A rancid taste characterizes stale-
ness. The fixed or expressed oil, both that of the sweet and
that of the bitter almonds (which are identical in taste, odor,
and other properties), is used in perfumery for fine hair oils,
ointments, and some fine soft soaps.

SWEET-FLAG ROOT.

Latin—Radix Calami; *French*—Racine de glaïeule; *German*—
Calmuswurzel.

The calamus root met with in commerce is the creeping
root-stock of a plant (Acorus Calamus), occurring in all countries of the northern hemisphere, and frequent in European
and American swamps. The root-stock is spongy, about as
thick as a finger, many-jointed, and of a yellowish color, with
many dark streaks and dots. Inside the color is reddish-white.
The odor is strong and the taste sharp and burning.

SWEET-PEA.

Latin—Lathyrus tuberosus; *French*—Pois de senteur; *German*
—Platterbsenblüthen.

Natural sweet-pea oil is unknown in commerce, but there are
specialties of the same name on the market, some of which repro-

duce the flowery character of the natural odor with great fidelity while others resemble it to some extent but lack floralcy.

Syringa.

Latin — Philadelphus coronarius; *French*—Seringat, Lilac; *German*—Pfeifenstrauchblüthen.

The white flowers of this garden bush have a very pleasant odor which resembles that of orange flowers, in place of which it can be used, in the cheaper grades of perfumery. This plant which flourishes freely in our climate deserves more attention by perfumers than it has hitherto received, since it appears to furnish an excellent substitute for the expensive oil of orange flowers, as above stated, in cheap perfumes.

Thyme.

Latin—Thymus Serpyllum; *French*—Thym; *German*—Thymian.

This well-known aromatic plant, which grows most luxuriantly on a calcareous soil, has an odor which is not unpleasant but is in greater demand for liqueurs than for perfumes. Here and there, however, it is employed for scenting soap. Common thyme, Thymus vulgaris, is used for the same purposes.

Under the name of Oil of Thyme, in the English and American market, is generally understood the oil of Thymus vulgaris, which is largely distilled in the South of France.

Tolu Balsam.

Latin—Balsamum tolutanum; *French*—Beaume de Tolu; *German*—Tolubalsam.

This balsam is derived from a tree indigenous to the northern portion of South America, Toluifera Balsamum, belonging to the Order of Leguminosæ. The balsam, which is obtained by incisions into the bark of these trees, is at first fluid, but becomes firm in the air owing to rapid resinification; in commerce it appears in a viscid form ranging from that of Venice turpentine to that of colophony. Its color varies from honey-yellow to reddish-brown; the taste is at first sweet, then sharp; it softens under the heat of the hand, and when warmed or sprinkled in powder form on glowing coals it diffuses a very pleasant odor recalling that of Peru balsam or vanilla. It shares with storax and Peru balsam the valuable property of fixing volatile odors and is often employed for this purpose, but is also frequently used alone in fumigating powders, tooth powders, etc. Adulteration of Tolu balsam with Venice turpentine or colophony is not rarely met with.

TONKA BEANS.

Latin—Fabæ Tonkæ; *French*—Fèves de Tonka; *German*—Tonkabohnen, Tonkasamen.

The South American tonka tree, Dipteryx odorata, bears almond-shaped drupes almost as long as the finger, which contain seeds two to four centimetres in length, the so-called tonka beans. These occur in European commerce in two sorts, the so-called Dutch and English tonka beans; the former are large, full, covered externally with a folded brown to black skin, and white inside. The latter are barely two-thirds the size of the former, almost black, and less glossy. The odor of the tonka bean is due to a volatile crystalline substance, coumarin, which often lies on the surface and in the wrinkles of the bean in the form of delicate, brilliant crystalline needles. Coumarin exists also in many other plants, for instance, in sweet woodruff (Asperula odorata), deer-tongue (Liatris odoratissima), etc.

TUBEROSE.

Latin—Polianthus tuberosa; *French*—Tubérose; *German*—
Tuberose.

The unusually strong and heady odor of the tuberose is unobtainable by distillation and was formerly extracted only by the
enfleurage process. The discovery of the volatile solvent process
and its refinement by Dr. Eugene Charabot permits the preparation of this valuable odor in the more convenient forms of the
floressence and hyperessence.

VANILLA.

Latin—Vanilla aromatica, Vanilla planifolia; *French*—Vanille;
German—Vanille.

The vanilla, which may justly be called a king among aromatic plants, is a climbing orchid indigenous to tropical America. It is cultivated on a most extensive scale on the islands
of Reunion and Mauritius; largely also in Mexico, and in
some other countries. The agreeable odor is present in the
fruit. These form three-lobed capsules about the length of
a lead pencil and the thickness of a quill. Externally they
are glossy brown, have a fatty feel, and show in the depression
a white powder which appears crystalline under a lens. Internally good fresh vanilla is so oily that it stains the fingers
on being crushed and is filled with numerous shining seeds
the size of a small pin's head. These properties, together
with the plump appearance and great weight, mark good qualities. Old vanilla, whose odor is fainter and less fragrant,
may be recognized by its wrinkled surface, the absence of the
white dust, the slight weight, and the bent ends of the capsules. - Fraudulent dealers endeavor to give such old goods
a fresher appearance by coating them with almond oil or Peru

balsam. "Vanilla de Leg" is recognized as the first quality of Mexican vanilla. Like most odors, that of vanilla does not become pleasant until it is sufficiently diluted.

VERBENA.

Latin—Verbena triphylla, Aloysia citriodora; *French*—Verveine; *German*—Verbenakraut.

The leaves of this Peruvian plant, especially on being rubbed between the fingers, exhale a very pleasant odor which is due to an·essential oil. The odor resembles that of fine citrons, or rather that of lemon grass; hence these two odors are frequently mistaken for each other. Owing to the high price of true oil of verbena,   the perfumes sold under this name are prepared from oil of lemon grass (see under Citronella) and other essential oils.

VETIVER.

Latin—Andropogon muricatus; *French*—Vétyver; *German*—Vetiverwurzel.

Vetiver, also called cuscus, and sometimes iwarankusa (though this is more properly the name of Andropogon lanifer; see above, under Citronella), is the fibrous root-stock of a grass indigenous to India, where fragrant mats are woven from it. The odor of the root somewhat resembles that of santal wood, and is used partly alone, partly for fixing volatile perfumes. Shavings of the root are frequently employed for filling sachet bags.

VIOLET.

Latin—Viola odorata; *French*— Violette; *German*—Veilchen blüthen.

The wonderful fragrance of the March violet is due to an essential oil which it is, however, difficult to extract. For this reason genuine perfume of violets, really prepared from the flowers, is among the most expensive odors, and the high-priced so-called violet perfumes are generally mixtures of other fine odors, while the cheaper grades are made from orris.

The manufacture of the synthetic ionones and the more flowery and therefore more valuable violettones now permits the perfumer to attain satisfactory results without the use of more than minute proportions of the very expensive natural violet oils, which, however, are of the greatest value in adding the final touch of true violet floralcy to the finished compositions. The violet character of orris root renders it of the greatest importance, which is heightened by the fact that this useful product is now available in the convenient form of the resinarome, a concentrated and completely soluble liquid extract of the odorant principles of orris root.

WALLFLOWER.

Latin—Cheiranthus Cheiri; *French*—Giroflé; *German*—Levkojenblüthen, Goldlack.

The wallflower, a well-known biennial garden plant belonging to the Order of Cruciferæ, according to recent experiments yields a very fine odor to certain substances and may be employed in the manufacture of quite superior perfumes. The preparations usually sold as wallflower, however, are not made from the flowers of this plant, but are mixtures of different odors.

WINTERGREEN.

Latin—Gaultheria procumbens; *French*—Gaulthérie; *German*—Wintergrünblätter.

This herbaceous plant, indigenous to North America, especially Canada and the Northern and Middle United States, where it grows wild in large quantities, has a very pleasant odor due to an essential oil and a compound ether which can also be produced artificially. The oil of wintergreen serves chiefly for scenting fine soaps.

YLANG-YLANG.

Oil Ylang-Ylang is obtained from the flowers of Uona odoratissima, a plant indigenous to the Philippine Islands. The culture has spread to the Bourbon Islands and to Madagascar, but Philippine Oil Ylang-Ylang is still in much greater demand and brings considerably higher prices than the ·Bourbon and Madagascar varieties. The oil is exceedingly fragrant and a valuable ingredient in many odors if of high quality. As is the case with Oil Patchouly, the quantities used are such that there should be little temptation for the perfumer to employ any but the best, though good results can frequently be obtained by substituting the best grades of Bourbon or Madagascar for the more expensive product from the Philippines. Philippine Ylang-Ylang is not produced in large amounts and under normal conditions the superior grade is consistently scarce.

The information given in the foregoing regarding the sources of the various perfume substances is mainly of value in informing the perfumer as to the relative quality of the oils from different localities. In the past perfumers were frequently compelled to secure the aromatic plants or flowers and extract the odorant principles themselves. This is never done at present except in very rare instances, as the oils are available in commerce in adequate amounts and in excellent quality, provided the user will take the necessary care to choose the best and to buy them from a firm whose reputation guarantees the purity of the products handled. Picking up bargains in perfume materials is rarely satisfactory, for, except in a few infrequent cases, these products must always be sold at prices commensurate with their quality, and comparatively low prices will be found nearly always to be synonymous with correspondingly low quality.

The list of aromatic plants here enumerated is necessarily incomplete, nor is it improbable that further refinements in the art of extracting the odorant principles will make possible the preparation of essential oils from many of the plants and flowers which have hitherto proved impractical as sou.ces of perfumes. New plants and flowers may also be added to the list through exploration of the tropical regions and through closer investigation of the resources of better known localities, but too much should not be expected. . The present natural odorant substances used in connection with the steadily growing class of synthetic perfume materials make possible the production of an infinite variety of odors. In making this statement the word infinite is used in the literal sense; mathematics rebels at the task of calculating the probable number of odors which can be made from the materials now in hand. But not all, by any means, of these odors are useful or attractive; despite the myriad of possibilities the task of the perfumer who sets out to originate an odor which will be sufficiently different to attain distinctiveness and sufficiently attractive to arrive at extensive popularity is amazingly difficult.

CHAPTER V.
THE ANIMAL SUBSTANCES USED IN PERFUMERY.

WHILE the vegetable kingdom offers us an abundance of aromatic odors the end of which it is impossible to foresee, the animal kingdom contains absolutely no substance which may be called sweet-scented in the strict sense of the term. If we find nevertheless a few animal substances generally used in perfumery, they should be considered rather as excellent means for fixing subtle vegetable odors than as fragrant bodies in the true sense. By themselves, indeed, they have an odor, but to most persons it is not agreeable even if properly diluted. Thus far only four substances of animal origin are employed in perfumery, namely: ambergris, castor, musk, and civet.

AMBERGRIS.

Latin—Ambra grisea; *French*—Ambregris; *German*—Ambra.

This is a substance whose origin is still doubtful; many facts indicate that it is a secretion—whether normal or morbid may be left undecided—of the largest living mammal, namely, of the pot-whale (Physeter macrocephalus). Ambergris is found in the intestines of this animal or, more frequently, floating about in the sea; the shores of the continents bordering the Indian Ocean furnish the largest amount of this peculiar substance.

Ambergris is a grayish-white fatty substance which occurs in commerce in pieces of various sizes—those as large as a fist are rare—of a penetrating, decidedly disagreeable odor. It is soluble in alcohol, and when properly diluted the odor becomes pleasant and it is so permanent that a piece of linen moistened with it smells of it even after being washed with soap. By itself, ambergris is not much used; it finds its chief application in combination with other odors or as an addition to some perfumes in order to make them lasting.

CASTOR.

Latin—Castoreum; *French*—Castoreum; *German*—Castoreum.

This is a secretion of the beaver (Castor fiber); it accumulates in two pear-shaped bags on the abdomen of the animal, both male and female. The hunters remove these bags from the body of the dead animal and in this form they are brought into commerce. These sacs are the length of a finger, at the thickest point the diameter of a thumb, and contain a greasy mass of yellowish-brown, reddish-brown, or blackish color, according to the nourishment of the animal. This mass constitutes castor; it has a strong, disagreeable odor, a bitter, balsamic taste, becomes soft when heated, is combustible, and almost entirely soluble in alcohol. It is probable that this secretion in its composition has some relation to the nourishment of the beavers which feed by preference on resinous vegetable substances. In commerce Canadian and Siberian castor are distinguished; the latter is more valuable and has almost disappeared from the market. It possesses a peculiar tarry, Russian-leather odor, probably due to a substance present in birch bark, upon which the Siberian animals feed almost exclusively. Canadian castor has an odor more nearly resembling pine resin. In perfumery castor is rarely used, usually only for fixing other odors.

MUSK.

Latin—Moschus; *French*—Musc; *German*—Moschus.

Of animal substances, musk is most frequently used in perfumery, and possesses the most agreeable odor of them all. Moreover, the odor of musk is the most intense that we know, actually imponderable quantities of it being sufficient to impart to a large body of air the strong odor of musk. This substance is derived from a deer which attains the size of a small goat and, like the chamois of the Alps, lives on the highest mountains of the Himalayas. Only the male animal (Moschus moschiferus) produces musk, which is secreted in a sac or rather gland near the sexual organ. Musk being subject to the worst adulterations owing to its high price, we append a description of the substance as well as of the sac or bag in which it appears in commerce.

The musk bag cut by the hunter from the body of the animal has the size and shape of half a walnut. On the side by which it was attached to the body of the animal it is membranous and nearly smooth; on the external surface it is more or less hemispherical and covered with light brown or dark brown hair, according to the season at which the animal was killed. The hair assumes a circular arrangement around an opening situated in the centre of the bag. This opening, the efferent duct of the gland, is formed by a ring-shaped muscle which yields to the pressure of a pointed object and permits the introduction of the point of the finger. Internally the musk bag consists of several layers of membrane which surround the musk itself. It is probable that the musk is secreted by these membranes, for when the animal is dissected, no direct communication of the musk gland with the body can be detected.

It has been surmised that the secretion of musk bears some relation to the food; at least it has been asserted that the animals eat, among other things, sumbul root with great avidity ; and this root, it will be remembered, has a .very intense odor of musk.   However, though this appears probable at first sight, it is contradicted by the fact that the females and the young males likewise eat the root without manifesting any odor of musk nor do they secrete the substance, while the older males produce it even when they are fed with hay only.   Another fact is of interest, namely, that other ruminants, too, for instance, cattle, diffuse a marked though faint odor of musk which occurs also in their excrements, exactly as in the case of the musk deer.   Alligators likewise produce a musk-like substance which has actually been made use of in place of musk for coarser purposes.

The musk present in the glands differs in appearance with the season and the age of the animal. Musk deers killed in spring have in their musk bag an unctuous soft mass of a reddish-brown color with the strongest odor; at other seasons the mass is darker in color, almost black, and granular; the size of the grains ranges from that of a millet-seed to that of a large pea.

That the secretion of musk belongs to the sexual functions appears probable from the fact that it can be found only in the bags of males more than two years old; that of younger animals contains only a substance of a milky consistence, whose odor has no resemblance to that of musk. The quantity of musk present in a bag varies with the season and the age of the animal; the smallest quantity may be assumed at about six drachms, though some bags contain as much as one and a half ounces.

The hunters dry the bags either on hot stones or in the air, or they dip them into hot oil. In commerce musk occurs either in bags under the name moschus in vesicis, "musk in pods," or free, moschus in granis, moschus ex vesicis, "grain musk." According to its origin four sorts are distinguished: Chinese or Tonquin musk, Siberian or Russian musk, Assam or Bengal musk, and finally Bokharian musk. The latter two varieties, however, rarely reach this market. Chinese musk (Tonquin or Thibet musk) occurs in small boxes containing twenty to thirty bags, each wrapped in Chinese tissue paper, on which Chinese characters are printed. This is considered the best quality. Assam musk occurs in boxes lined with tin which contain as many as two hundred or more bags; its value is about two-thirds that of the former. Russian musk is packed in various ways and is worth about one-fourth that of the Chinese; a special variety of it, of a weaker and rather urinous odor, is known as Cabardine musk; of least value is Bokharian musk which is of a grayish black color, with a faint odor.

Musk is adulterated in an almost incredible manner; at times so-called musk bags are met with which are artificially constructed of animal membranes and filled with dried blood, earth, etc., and slightly scented with genuine musk. But even the genuine musk bags are often tampered with; musk being removed from the opening and the space filled with earth, dried blood, animal excrement, or perhaps pieces of copper and lead.

Pure musk reacts quite characteristically toward caustic

alkalies such as caustic potash and soda or solution of ammonia, and these substances are used for testing the purity of musk. If a dilute alkaline solution is poured over musk, a marked increase of the odor is observed after a short time; if the alkaline solution is concentrated or hot, the odor of musk disappears completely and the fluid develops the caustic odor of pure ammonia. Hot water dissolves about eighty per cent of the total weight of musk; strong alcohol dissolves about one-tenth of it; when heated in an open porcelain capsule, musk burns with a disgusting empyreumatic odor and leaves a considerable amount of ash, about one-tenth of its weight. Besides the above-named substances which destroy the musk odor by the decomposition of the aromatic constituent, there are other bodies, whose action we do not know at present, which have the peculiar property of completely extinguishing this most penetrating of all odors: to deodorize a vessel completely which has contained musk, it is sufficient to rub in it some bitter almonds moistened with water or some camphor with alcohol.

In an extremely dilute condition musk is used for perfuming the finest soaps and sachets, and even in the manufacture of the most expensive and best perfumes, owing to its property of imparting permanence to very volatile odors. In the last-mentioned class, however, the quantity of musk must always be so small that its presence is not distinctly observed, since many persons find the pure odor of musk very disagreeable, while they praise the fragrance of such perfumes as contain an amount of this substance too small to be perceived by the olfactory nerves.

CIVET.

Latin—Civetta; *French*—Civette; *German*—Zibeth.

This substance bears some resemblance to musk with reference to its derivation and the rôle it plays in the life of

the animal from which it is obtained.   The Viverridæ, a class
of carnivora related to the cats and weasels, found in Asia
and Africa, furnish this substance.   It is obtained chiefly
from the civet cat (Viverra Civetta) and the musk rat (Viverra
Zibetha) which are kept in captivity for the purpose of ab-
stracting from them from time to time the civet which is
always formed anew.

Civet is the secretion of a double gland present both in
the male and the female near the sexual organs.   Fresh civet
is a whitish-yellow mass of the consistence of butter or fat,
and becomes thicker and darker on exposure to the air.
Similar to musk, it has a strong odor which becomes pleasant
on being diluted and is used both alone and for fixing other
odors.

CHAPTER VI.

THE CHEMICAL PRODUCTS USED IN PER-
FUMERY.

IN the manufacture of perfumery a considerable number
of chemical products find application; in this place, however,
we shall describe only those which are used very frequently
and generally, and discuss the characteristics of those employed more rarely in connection with the articles of perfumery into which they enter.   According to their application
we may divide these substances into several groups, namely:

A. Chemicals which, without themselves serving as perfumes, are used exclusively for the extraction of odors.

B. Chemicals which, while not fragrant, are frequently
employed in the preparation of perfumes.   Under this head
we have included also those substances which are not strictly
chemical products, but originally come from the animal or

vegetable kingdom, such as fats, spermaceti, and wax, yet cannot be used in perfumery, unless they have undergone a process of chemical purification.

C. Chemical products used for coloring perfumes, so-called due-stuffs.

The coloring of perfumed products is an important matter. Not only is there the comparatively simple task of coloring the extracts and toilet waters, but there is the most difficult problem of tinting face powders and rouges and of coloring the many shades of toilet soaps. Vegetable dyes are used to some extent, but for the most part dependence is placed on the artificial dye-stuffs which are available in such a wide variety of shade and chemical character as to permit the attainment of any desired effect.

Few if any of the substances to be treated here will ever be prepared by the perfumer, but a knowledge of their manufacture and characteristics is valuable if not absolutely essential.

A. CHEMICALS USED FOR THE EXTRACTION OF AROMATIC SUBSTANCES.

For the extraction of aromatic substances from plants a number of bodies are used which possess great solvent power for essential oils, and are besides very volatile, or have a low boiling-point. These are particularly ether, chloroform, petroleum ether, and bisulphide of carbon.

ETHER.

This liquid, in commerce also called sulphuric ether, is made in large quantities in chemical laboratories by the distillation of alcohol with sulphuric acid, followed by a second distillation or rectification. When pure, ether forms a mobile, thin, strong-smelling, and inflammable liquid which when inhaled produces insensibility, for which reason it is used as an anæsthetic in surgery. Its specific gravity is about 0.720 when anhydrous, and its boiling point 35° C. (95° F.). It forms an excellent solvent for essential oils, resins, fats, and similar bodies.   Owing to its great volatility, its vapors are quickly diffused in the air, and, as they are very inflammable, lights must be kept away from a bottle containing this substance.   The same remark applies to most of the substances to be presently described.

CHLOROFORM

is prepared by the distillation of chlorinated lime, alcohol, and water, acetone being more recently substituted for the alcohol, followed by rectification of the product.   When inhaled it produces insensibility like ether.   It has a pleasant odor and sweet taste.   Its specific gravity is about 1.49 and its boiling-point 61° C. (142° F.).   Owing to its great solvent power and low boiling-point, chloroform is largely used for the extraction of aromatic vegetable substances; it does not take fire directly in the air.

PETROLEUM ETHER.

Petroleum, which is brought into commerce in immense quantities, especially from Pennsylvania, for illuminating purposes, cannot be used in its crude state, but requires rectification.   Petroleum as it issues from the earth consists of various hydrocarbons mixed together, some of which have very low boiling-points, so that their vapors readily take fire and would make the use of petroleum in lamps dangerous.   Petroleum, therefore, is heated in large apparatuses to about 70 or 80° C. (158 to 176° F.), when the more volatile products pass over, and the petroleum for illuminating purposes remains in the stills.   A certain fraction of the volatile distillate, the so-called petroleum ether, is largely used in the manufacture of varnishes.   Owing to its great solvent power for aromatic vegetable substances and its low price, petroleum ether has become quite an important body for the extraction of perfumes, which will be further discussed hereafter.   Good pe-

5

troleum ether is colorless, has a peculiar, not unpleasant odor, and a boiling-point between 50 and 55° C. (112° and 131° F.).

BENZIN

is a common name for another fraction of the volatile distillate from petroleum, viz., that which boils between 50° and 60° C. (122° to 140° F.) and has a spec. grav. of 0.670 to 0.675°.

This liquid, which is also used as a volatile solvent for the extraction of odorous substances, must not be confounded with Benzene or Benzol, a distillate from coal tar, boiling at about 80° C. (176° F.) and having a spec. grav. of 0.878. The latter is not used for the extraction of perfumes.

BISULPHIDE OF CARBON.

This is made by conducting vapors of sulphur over glowing charcoal or coke. The vapors of bisulphide of carbon thus formed are led into vessels filled with ice or ice-cold water, where they condense. Bisulphide of carbon is a colorless liquid, heavier than water and very refractive. It is inflammable, and possesses a peculiar odor which is not disagreeable if the liquid has been thoroughly purified. Its boiling-point is about 45° C. (113° F.) and it has great solvent power. At the present time, the market affords bisulphide of carbon of a high degree of purity.

Some manufacturers who prepare their odors by extraction, may find it advantageous to make also the bisulphide of carbon necessary for it, and this is best done in Gérard's apparatus (Fig. 1). It consists of a cast-iron cylinder *a*, two metres high and one metre in diameter. This cylinder is heated on the outer surface in an oven, and two tubes, *c* and *d*, are attached to it. Tube *d* is connected by *e* with the hemispherical vessel *b* which is connected by the tube *i* with the condenser *mlk*. The condenser is formed of three cylinders made of sheet zinc which are surrounded with cold water. The condensed liquid escapes into the vessel p, while the gaseous products pass through n into the chimney. The cylinder a is filled with about 1,500 pounds of charcoal or coke in small pieces, after which it is closed and all tubes are carefully luted with clay; a is then heated to a strong red heat and at intervals of three minutes 3 pounds of sulphur are thrown in through c. In twenty-four hours, by the use of 478 pounds of sulphur, 568 pounds of crude bisulphide of carbon are obtained; a

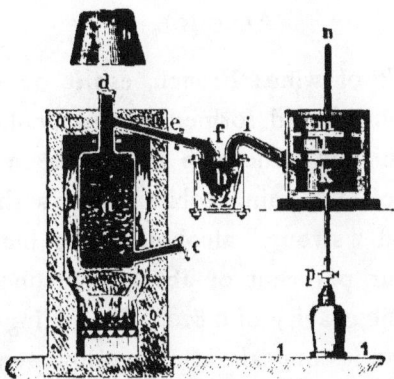

FIG. 1.

portion of the sulphur distils over uncombined into the vessel b.

The crude bisulphide of carbon contains about twelve per cent of sulphur and other combinations in solution and is redistilled at exactly 48° C. (118.4° F.) in a steam-heated apparatus with a long exit tube cooled with ice below and water above. In order to obtain the bisulphide of carbon absolutely pure, which is essential to render it suitable for extraction, it is again distilled at the same temperature, with the addition of two per cent of palm oil. As the vapors of bisulphide of carbon are injurious to the organism, the vessels containing it must always be kept well closed.

B. Chemical Products used for the Preparation of Perfumes.

Among all the substances belonging under this head, there is one which plays a prominent part in the manufacture of most perfumes.   In handkerchief perfumes it is one of the most important substances, as it forms not only the greatest bulk, but the perfection of the perfume depends upon its quality.   This substance is—

ALCOHOL,

also called spirit of wine; French, esprit de vin; the well-known combustible liquid formed by the alcoholic fermentation of sugar, which is made on a large scale in extensive distilleries.   Alcohol is a thin, mobile liquid with an aromatic odor.   The usual " strong " alcohol of the market contains about ninety-four per cent of absolute alcohol by volume. This has a specific gravity of 0.820. Its boiling-point is 78.2° C. (172.4° F.), and it congeals at a very low temperature, below −100° C.   Alcohol possesses great solvent power for resins, balsams, and essential oils.

These properties, however, belong only to the commercial stronger or so-called " druggists' alcohol," and more particularly to a very pure quality of it, as free as possible from fusel-oil compounds, known as cologne spirit.   As absolute alcohol is also necessary for the purposes of perfumery, we shall briefly describe its preparation.

In order to make absolute alcohol, sulphate of copper is heated in a retort until it has changed into a white powder. After the powder has cooled in the covered retort, it is at once introduced into a large glass bottle; over it is poured the strongest obtainable alcohol (96% Tralles) which must be free from fusel oil; then the bottle is closed air-tight and repeatedly shaken. The sulphate of copper which has lost its water of crystallization by the heat reabsorbs it from the alcohol and again becomes blue and crystalline. Generally four pounds of sulphate of copper are used for ten quarts of alcohol; when white burnt sulphate of copper after long contact with alcohol still remains white, the alcohol is proved to be practically anhydrous (it may still contain about two per cent of water).

Larger quantities of absolute alcohol are made in a copper still containing fused anhydrous chloride of calcium in small pieces. The apparatus is closed and alcohol of 94 to 95% is poured in through a tubulure. The mixture often grows so warm that the alcohol begins to pass over, so that but little heat need be applied to make the absolute alcohol distil over.

Absolute alcohol obtained in this way—for by repeated distillation we get at most an alcohol of 96%—abstracts water from the air with avidity; hence it must be preserved in airtight vessels which should contain a small amount of anhydrous sulphate of copper.

Strong commercial alcohol contains varying amounts of water—from four to twenty parts by volume (96 to 80% alcohol); at the present time, however, it is always customary for dealers in this country to supply the officinal alcohol of 94%, when "strong alcohol" is called for. Its strength is measured by an areometer which sinks in proportion to the purity of the alcohol; the alcoholometer of Tralles or volumeter shows at once on its scale how many parts by volume of absolute alcohol (volume per cent) are contained in 100

FIG. 2.

volumes of alcohol. The adjoining figure (Fig. 2) shows Tralles' alcoholometer, with the vessel in which the test is

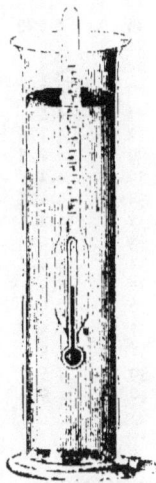

made. The readings of the instrument, however, are correct only at a temperature of 15.6° C. (60° F.), the so-called normal temperature; at a higher or lower point they must be corrected according to the tables appended.

At temperatures below the normal, the amount of alcohol is greater than the areometer indicates, hence a percentage must be added; at higher temperatures a percentage must be deducted.

TABLES FOR FINDING THE TRUE PERCENTAGE BY VOLUME, AT THE NORMAL TEMPERATURE OF 60° F., OF ALCOHOL OF ANY STRENGTH, WHEN TESTED AT TEMPERATURES BELOW OR ABOVE 60° F.

TABLE I.—FOR TEMPERATURES UNDER 60° F.

Per cent of Alcohol by Volume.	Number of F. Degrees Requiring Addition of one to Percentage.	Per cent of Alcohol by Volume.	Number of F. Degrees Requiring Addition of one to Percentage.	Per cent of Alcohol by Volume	Number of F. Degrees Requiring Addition of one to Percentage.	Per cent of Alcohol by Volume.	Number of F. Degrees Requiring Addition of one to Percentage.
21	5.4	41	4.725	60	5.4	79	6.3
22	5.175	42	4.725	61	5.4	80	6.3
23	4.725	43	4.725	62	5.4	81	6.525
24	4.5	44	4.725	63	5.625	82	6.525
25	4.5	45	4.95	64	5.625	83	6.75
26	4.5	46	4.95	65	5.625	84	6.75
27	4.5	47	4.95	66	5.625	85	6.75
28	4.275	48	4.95	67	5.625	86	6.75
29	4.275	49	4.95	68	5.85	87	6.975
30	4.275	50	5.175	69	5.85	88	7.2
31	4.275	51	5.175	70	5.85	89	7.425
32	4.275	52	5.175	71	5.85	90	7.65
33	4.275	53	5.175	72	5.85	91	7.875
34	4.275	54	5.175	73	5.85	92	8.1
35	4.5	55	5.175	74	6.075	93	8.325
36	4.5	56	5.175	75	6.075	94	8.775
37	4.5	57	5.4	76	6.075	95	9.
38	4.5	58	5.4	77	6.075	96	9.45
39	4.5	59	5.4	78	6.3	97	10.125
40	4.5						

EXPLANATION.—Supposing an alcohol should be found to contain 40 per cent of absolute alcohol by Tralles' alcoholometer at 45° F. The difference between 45 and 60° F. is 15. Opposite to 40 will be found the figure 4.5. For every 4.5 degrees F. below 60° there must be added 1 to the alcoholic percentage. Hence for 15 degrees there must be added 3.3 degrees. The alcoholic percentage, by volume, therefore, is 43.3 per cent.

TABLE II.—FOR TEMPERATURES ABOVE 60° F.

Per cent of Alcohol by Volume.	Number of F. Degrees Requiring SUBTRACTION of one from Percentage.	Per cent of Alcohol by Volume.	Number of F. Degrees Requiring SUBTRACTION of one from Percentage.	Per cent of Alcohol by Volume.	Number of F. Degrees Requiring SUBTRACTION of one from Percentage.	Per cent of Alcohol by Volume.	Number of F. Degrees Requiring SUBTRACTION of one from Percentage.
21	5.85	41	4.5	61	5.175	81	6.075
22	5.625	42	4.5	62	5.175	82	6.075
23	5.4	43	4.5	63	5.175	83	6.3
24	5.175	44	4.5	64	5.175	84	6.3
25	4.95	45	4.5	65	5.175	85	6.3
26	4.95	46	4.5	66	5.4	86	6.525
27	4.725	47	4.725	67	5.4	87	6.525
28	4.725	48	4.725	68	5.4	88	6.525
29	4.5	49	4.725	69	5.625	89	6.75
30	4.5	50	4.725	70	5.625	90	6.975
31	4.5	51	4.725	71	5.625	91	6.975
32	4.5	52	4.725	72	5.625	92	7.425
33	4.5	53	4.95	73	5.625	93	7.425
34	4.5	44	4.95	74	5.625	94	7.65
35	4.5	55	4.95	75	5.85	95	7.65
36	4.5	56	5.175	76	5.85	96	8.1
37	4.5	57	5.175	77	5.85	97	8.1
38	4.5	58	5.175	78	5.85	98	8.325
39	4.5	59	5.175	79	6.075	99	9.45
40	4.5	60	5.175	80	6.075	100	9.9

EXPLANATION.—In this case, the same calculation is performed as directed under Table I., except that the correction is to be *deducted* instead of added.

Aside from the water present in it, commercial alcohol is never pure, but always contains small quantities, at times mere traces, of substances having a peculiar, sometimes pleasant, sometimes disagreeable, but invariably intense odor, which are known as fusel oils. The variety of fusel oil differs with the raw material from which the alcohol was made; there is a potato fusel oil (chemically amyl alcohol), a corn fusel oil, a beet fusel oil, wine fusel oil (œnanthic ether), etc. Fusel oils, being themselves odorous substances, exert an influence on the fragrance of the perfume; hence it is a general rule in perfumery to use only alcohol free from fusel oil; that is, such from which the fusel oil has been extracted as far as possible by means of fresh charcoal. So-called "Cologne Spirit" of the best quality is, as a rule, practically free from it.

Strange to say, some essential oils or aromatic substances in general, develop their finest odors only when the perfumes are prepared with an alcohol from a certain source. While the charcoal treatment removes almost all the fusel oil, the remaining traces suffice to act as odorous substances in the true sense of the term and to produce with other aromatic bodies a harmony of the odor which can never be reached by the use of another variety of alcohol. To give but a single instance we may state that all the citron odors known in perfumery develop the finest aroma only when dissolved in alcohol made from wine and the solution is then distilled. The world-renowned eau de Cologne is made in this way; the other aromatic substances contained in it are added to the distillate from the spirit of wine and the citron oils; any cologne made in another manner or with another alcohol has a less fine odor. While the citron odors require true spirit of wine for the development of their full aroma, other scents require beet or corn alcohol to bring out their best odor. Jasmine, tuberose, orange flowers, violet, etc., and all animal odors (ambergris, musk, and civet) belong to the latter class. For this remarkable and to the perfumer most important fact we know no other explanation than that traces of fusel oils present even in rectified alcohol take part in the general impression made on the olfactory nerves, acting as true aromatic substances.

Cologne spirit is expensive, but this should not be a reason for accepting a cheaper grade, with which it would be absolutely impossible to make really fine perfumes.

Alcohol is also generally used for the direct extraction of odorous substances from plants, as will be seen in the description of the processes employed in the preparation of the so-called essences or extracts. For these purposes, too, the best cologne spirit only should be used, that is, alcohol which has been freed from fusel oil and redistilled, for in no other way can the aromatic substances be obtained in the greatest possible purity. And this is indispensable for the preparation of really fine perfumes, for we do not hesitate to say that French and English perfumes have acquired their deserved reputation mainly through the great care exercised in the selection of their raw materials, and especially of the alcohol used for extraction.

ALLOXAN.

This preparation, which is used in making a fine skin cosmetic, is manufactured in chemical laboratories from uric acid heated with nitric acid. Alloxan is a crystalline colorless body which has the property of gradually producing a red tint on the skin and finds employment for this reason.

AMMONIA.

Ammonia is a gas formed by the decomposition of nitrogenous substances, but chiefly obtained, on a large scale, from the so-called "gas liquor" of gas works. By itself it develops a very disagreeable odor and stimulates the lachrymal glands to secretion—a fact which can be verified in any stable. A solution of the gas (water of ammonia; liquor ammoniæ) possesses the same properties. In perfumery ammonia is never used alone, but only in combination with other odors, namely, in the manufacture of smelling salts (French: sels volatils; German: Riechsalze), which are much in favor in England and in this country. For the purposes of the perfumer, the greater part of the commercial ammonia is unsuitable owing to its tarry odor. Pure ammonia is best prepared by heating equal parts of quicklime and powdered sal-ammoniac in a retort, and conducting the generated gas into water which dissolves it with avidity, one quart of water dissolving more than seven hundred quarts of ammonia gas.

CARBONATE OF AMMONIA,

a combination of ammonia with carbonic acid, occurs in commerce in large transparent lumps, often covered with a white dust of bicarbonate of ammonia, which in the air continually develop ammonia and therefore always smell of it. This commercial product is, as a rule, sufficiently pure to be used in perfumery; as to its application the same remarks apply as were made under the head of ammonia.

OIL OF BITTER ALMONDS (OLEUM AMYGDALÆ AMARÆ).

This is made from bitter almonds, previously deprived of fatty oil by pressure, which are mixed with an equal weight of water and set in a warm place. The amygdalin undergoes decomposition into sugar, hydrogen cyanide, and benzoyl hydride or oil of bitter almonds. After one or two days the mass is distilled; the distillate being a colorless liquid, containing, besides oil of bitter almonds, hydrogen cyanide or prussic acid, one of the most virulent poisons, from which it must be freed. This is done by shaking the liquid repeatedly with dilute solution of potassa, followed by agitation with water. Pure oil of bitter almonds is not poisonous, but has a very strong narcotic odor of bitter almonds, which, however, becomes most marked when largely diluted with water.

BENZOIC ACID (ACIDUM BENZOICUM).

This acid, contained in benzoin, is made also synthetically from other materials, in chemical laboratories. When pure it forms needle-shaped crystals having a silky gloss; they have a peculiar acrid taste, but no odor. Synthetic benzoic acid is worthless to the perfumer; in his art he can use only a benzoic acid made from gum benzoin by sublimation, because it contains a very aromatic essential oil for which the acid is merely the vehicle and which can also be employed alone.

As this sublimed benzoic acid is often adulterated with the artificial, we advise the manufacturer of perfumery to make his own benzoic acid according to the following directions.

The Manufacture of Sublimed Benzoic Acid.

About four pounds of benzoin B of best quality is broken into small pieces and placed in a small copper boiler K (Fig. 3); over its entire surface is pasted white blotting paper L, and to this is pasted a cone of strong pa-
per which must surround the edge of the boiler. The cone ends above in a paper tube R, about five feet long and an inch wide. The copper boiler is placed in a large clay pot T (a flower pot) and sur-rounded on all sides with fine sand. The clay pot is heated from without by a char-coal fire. After the pot has remained about half an hour on the fire, the latter is fanned

FIG. 3.

to its utmost and kept at this point for thirty minutes. The heat volatilizes the benzoic acid, the above-mentioned essen-tial oil, and some tarry substances of a brown color. The latter are arrested by the filter paper, while the benzoic acid is deposited on the cone and in the tube, in the form of deli-cate glossy needles which are very fragrant owing to the essential oil. The largest yield of benzoic acid is obtained when the temperature is raised very gradually, until finally nothing remains in the copper boiler but a brown, almost car-bonized mass of a blistered appearance.

BORAX (SODII BORAS)

is used in some preparations. Borax forms colorless crystals which slightly effloresce in dry air and hence must be preserved in tightly closed vessels. Reddish tinted cystals are contaminated with oxide of iron and should be rejected.

PERMANGANATE OF POTASSIUM (POTASSI PERMANGANAS)

is a salt formed by fusing a mixture of manganese dioxide, potassa, and potassium chlorate, extracting the product with water, and evaporating the solution to crystallization; the salt is obtained in small dark violet, almost black crystals which dissolve in sixteen parts of water to which they impart a beautiful violet color. By contact with organic substances, or others easily oxidized, the solution changes its color into green and finally is decolorized, precipitating a brown powder. Owing to this change of color the salt has been called chameleon mineral. As its preparation requires considerable dexterity, it is preferable to buy it from reputable houses, rather than to make it. It is used in the manufacture of mouth washes and hair dyes. The solution of the salt causes brown stains on linen and the skin; they can be removed only if the spots are immediately washed with hydrochloric, oxalic, sulphuric, or another acid.

ACETIC ACID (ACIDUM ACETICUM).

Much confusion exists in the literature regarding the strength of acetic acid when merely called by this name. It is safe to assume that, in each country, the term applies to the acid officinal in its national pharmacopœia as " Acidum Aceticum." Thus the Austrian and German pharmacopœias understand by it an acid containing 96% of absolute acetic acid, which is practically identical with what is known as glacial acetic acid. The latter is, in some pharmacopœias, distinguished by a special name: acidum aceticum glaciale, U. S. P.; acide acétique crystallisable, French Pharm.—In the present work, the author always intended the strong acid of the Austrian pharmacopœia to be understood when no other strength was designated. Like alcohol, strong acetic acid dissolves essential oils and is used in the manufacture of various toilet vinegars and washes. Acetic acid is made in chemical laboratories by distillation of acetate of sodium with sulphuric acid, or more commonly from wood vinegar. The buyer should always satisfy himself that the product is free from an empyreumatic odor which clings tenaciously to an insufficiently purified sample.

FATS.

Fats find extensive application in perfumery, in the preparation of the so-called huiles antiques, pomades, and many other cosmetics. They should be enumerated among the chemical products used in perfumery because they can never be employed in their commercial form, but must undergo some process of purification, which is effected less by mechanical than by chemical means. Commercial fats usually contain remnants of the animal or vegetable body from which they are derived: particles of blood and membranes occur frequently in animal fats; cell bodies and vegetable albumin in vegetable fats. Besides these mechanical impurities, fats, especially if old, sometimes contain small amounts of free fatty acids which suffice to impart to them the objectionable odor and taste peculiar to every rancid fat. While some fats, such as bear's grease, butter of cacao, oil of sesame, and some others, remain free from rancidity for a long time, others undergo this change very rapidly; in fact, we may say that every fat which shows the slightest odor should be called rancid, for pure fat is absolutely odorless.

We shall here briefly describe the process employed in the fat industry and by perfumers for the purification of fats. Animal fat, such as lard, suet, bear's grease, etc., as well as cocoanut and palm oils, are introduced into a large iron boiler containing dilute soda lye (not exceeding one per cent of caustic soda), and the lye is heated to boiling. In the boiler is a small pump terminating above in a curved tube having a rose of a watering-pot at the end. The pump is so arranged as to raise lye and melted fat at the same time and to return the fluid into the boiler in a fine spray. After the fat is melted, the solid matters floating on top are skimmed off with a perforated spoon, and then the pump is operated for about fifteen minutes. The contained shreds of membrane and similar substances are completely dissolved by the soda lye, the free fatty acids are perfectly combined, and the fat is at the same time decolorized. After cooling, it floats on the surface of the lye as a colorless and odorless fluid; it is ladled off and poured into tall tapering vessels which are well closed and preserved in cool cellars. Contact with the air, especially at higher temperatures, causes rancidity of the fat. For every twenty pounds of fat twenty quarts of lye are used.

According to another process the fat is purified by being heated with alum and table salt; or every twenty-five pounds of fat, one ounce of alum and two ounces of salt are dissolved in five gallons of water. The scum is carefully skimmed from the surface of the melted fat, and, after it has solidified, the fat is washed with water until the latter escapes perfectly tasteless and odorless.

The washing is a very complicated and tedious piece of work. Operating on a small scale, a slightly inclined marble slab is taken, upon which a thin stream of water is constantly falling from a tube arranged above it. The fat is placed on the slab in small quantities (not over two pounds) and ground with a muller, like oil colors, under a constant flow of water. Owing to the expense of hand labor, it is advisable to use a so-called vertical mill or chaser. This consists of a level, circular, horizontal marble slab, bearing a central, easily movable axis with a crosspiece upon which two, likewise vertical, cylindrical marble plates turn like wheels in a circle on the horizontal marble plate. The fat is placed on the latter and continually irrigated with water; behind every chaser is applied a marble plate with a blade which nearly touches the chasers and returns the fat displaced laterally, under the chasers. The axis around which the chasers run is kept moving by any available power, and the laborer has nothing to do but to replace the washed fat with crude.

Liquid fats are purified as follows:

The oil is intimately mixed with one per cent of sulphuric acid. The mixture assumes a black color, the vegetable mucilage present in the oil becoming carbonized. After several days' rest the oil becomes clear and floats on the surface of the sulphuric acid which has assumed a black color from the presence of finely divided carbon. The oil is decanted and treated, in the manner above stated for solid fats, with caustic soda lye. Heating can be dispensed with if the pumping is continued for a longer time.

Benzoin and benzoic acid have the property of counteracting the tendency of fats to become rancid; it is advisable, therefore, to mix intimately with the completely washed fat a small amount of benzoic acid, at most one-one-thousandth part by weight.

The best way of preserving fats is by salicylic acid. This is added to solid fats while they are in a melted state; if oils, the acid is poured in and the bottle vigorously shaken. If the oil is in casks, a small bag filled with salicylic acid is hung into it from the bung-hole. The acid dissolves in the oil and is disseminated through it and thus effects its preservation. One-one-thousandth part by weight of the fat or oil is said to be more than sufficient to keep it perfectly fresh for years.

Fats differ largely in their physical properties—for instance, in their appearance, melting-point, firmness, etc. As we shall return to this subject in connection with the manufacture of some perfumes, it is enough here to state briefly that by the addition of spermaceti, wax, paraffin, etc., fats are made more transparent and firmer—a matter of importance for some cosmetic preparations.

CHINESE GELATIN.

This substance, derived from several algæ, species of Eucheuma, indigenous to the Chinese sea, and identical with Japanese agar-agar, on being boiled with two hundred parts of water has the property of forming a colorless solution which solidifies on cooling. Owing to this property the addition of a small quantity of Chinese gelatin (0.1-0.2%) is an excellent means for imparting to certain pomades and ointments great transparency and firmness.

FRUIT ETHERS

are liquids which possess an agreeable, refreshing odor resembling that of some fruits. For this reason they are used in confectionery, in the manufacture of liqueurs, and also in many ways in perfumery. Chemically, fruit ethers are combinations of an organic acid—acetic, butyric, valerianic, etc. —with a so-called alcohol radicle such as ethyl and amyl. Their manufacture is connected with many difficulties and is but rarely attempted by perfumers, especially as these products are made a specialty in some chemical laboratories and are furnished at very low prices and of excellent quality. In perfumery the following fruit ethers are particularly employed.

ACETIC ETHER,

prepared by the distillation of acetate of sodium with alcohol and sulphuric acid, is a colorless liquid having an odor of fermenting apple juice, with a boiling-point at 74° C. (155° F.).

PINE-APPLE ETHER

(ether or huile d'ananas) is made by the saponification of butter with solution of potassa, distillation of the soap with alcohol and sulphuric acid, and rectification of the distillate. It is an inflammable liquid with an intense odor of pine-apple; its boiling-point is 119° C. (246° F.). It is not generally used pure, as its odor needs some correction. This is accomplished by the addition of a little valerianate of amyl, and chloroform. Also in other ways.

APPLE ETHER,

prepared by distillation from valerianate of sodium with alcohol and sulphuric acid, and the subsequent addition of certain correctives (see below).

PEAR ETHER,

also called pear oil, chiefly valerianate of amyl oxide, can be obtained in large quantities from a by-product in the manufacture of potato spirit, namely, amyl alcohol, which is carefully heated in a still with bichromate of potassium and sulphuric acid. The product thus obtained has a very pleasant odor of fine pears and boils at 196° C. (385° F.). But the commercial "pear-essence" is a more complex body (see following table).

NITROUS ETHER

is a very volatile liquid boiling at 16° C. (61° F.), which is obtained by distillation of strong alcohol with concentrated nitric acid and rectification of the distillate; it is less used in perfumery than the other fruit ethers.

Fruit ethers, owing to their low price and great strength, are frequently employed in the manufacture of cheap perfumery, in place of essential oils, but more largely for scenting soap.

6

The so-called raspberry and strawberry ethers cons.st of mixtures of acetic, pine-apple, apple, and other ethers (see following table), which, combined in certain proportions, really manifest an odor nearly akin to those of the fruits after which they are named.

FRUIT ETHERS (FRUIT ESSENCES).

GLYCERIN.

This substance, which may be called a true cosmetic in itself, as it possesses marked solvent power for cutaneous coloring matters and at the same time imparts to the skin delicacy and flexibility, is at present to be had commercially in great purity. Pure glycerin is a brilliant, colorless, and odorless substance of the consistence of a thick syrup, which mixes with water and alcohol in all proportions and has a slightly warm but very sweet taste. It readily absorbs aromatic substances and is used in many valued toilet articles in combination with fats and perfumes. Recently we have succeeded in using glycerin most successfully for the extraction of aromatic substances.

OIL OF MIRBANE,

also called artificial oil of bitter almonds, nitrobenzol, and essence of mirbane. This substance, which is now largely used in perfumery and soap manufacture, is obtained by the action of fuming nitric acid on benzol. The mixture becomes hot and emits masses of brown vapors, and there is formed a yellow oily body which is washed with water and soda solution until the washings escape colorless. Pure nitrobenzol is not soluble in water, but in alcohol or ether, boils at 213° C. (415° F.), and congeals at −5 to 6° C. (21–23° F.). Its spec. grav. is 1.2 or a little over. Any oil of mirbane having a lower specific gravity than 1.2 at 15° C. (59° F.) is spurious, most likely nitrotoluol. Its odor greatly resembles that of oil of bitter almonds, but can be clearly differentiated from it on comparison. Care must be taken in inhaling the vapor when undiluted, as it is poisonous. By distillation nitrobenzol can be obtained quite colorless, and in this form is often used for the adulteration of genuine oil of bitter almonds. This adulteration, however, can be easily demonstrated by heating for a short time with an alcoholic solution of a caustic alkali which separates from nitrobenzol a brown resinous substance, while true oil of bitter almonds loses its odor and changes into benzoic acid which unites with the alkali.

PARAFFIN.

This substance is one of the products of the distillation of petroleum, coal, peat, and other carbonaceous sources. It is a crystalline, brittle body, closely resembling wax in appearance and melting between 51 and 60° C. (124 and 140° F.). Paraffin, which is now made on a large scale for the manufacture of candles, is very useful in perfumery as a partial substitute for the much more expensive wax or spermaceti, over which it has the advantage, besides its cheapness, that it imparts to the articles great transparency—a quality which is valued highly in fine perfumeries. The addition of some paraffin to pomades renders them more consistent and counteracts their tendency to become rancid. Distilled paraffin always has a crystalline form, differing from the paraffin-like residues left after the distillation of petroleum (so-called vaselins, etc., see below) which are always amorphous.

PRYROGALLIC ACID

appears in commerce as a white crystalline powder, made by heating gallic acid to 200–210° C. (392–410° F.). With iron salts, pyrogallic acid forms bluish-black combinations and precipitates the metal from silver solutions as a velvety-black powder. On account of these properties pyrogallic acid is used in perfumery as a constituent of some hair dyes.

SULPHIDE OF POTASSIUM,

liver of sulphur, hepar sulphuris, potassii sulphuretum, the pentasulphide of potassium, is obtained by fusing together potash and sulphur, in the shape of a leather-brown mass which is soluble in water and on exposure to the air is gradually decomposed with the development of the offensive sulphuretted hydrogen gas; hence it should be preserved in well-closed vessels. An aqueous solution of this substance forms with lead or silver salts a black precipitate of sulphide of lead or silver, and is used for some hair dyes.

STARCH FLOUR

(amylum) is prepared from various vegetables sucn as potatoes, rice, arrowroot, sago, etc., and when pure appears as an insoluble white powder which the microscope shows to be grains consisting of many superimposed layers. In commerce the price of the different varieties of starch fluctuates greatly; in perfumery well-cleansed potato starch can very well be used for dusting powders, and the so-called poudre de riz; in this country, corn starch is preferable.

VANILLIN,

that is, the body to which vanilla owes its fragrance, is now made artificially and can be used in place of vanilla.

VASELIN.

In the distillation of petroleum there remain in the still as a residue large quantities of a substance which when purified is colorless and, according to the nature of the petroleum, at ordinary temperatures has either the consistence of lard, melting under the heat of the hand, or forms an oily liquid. In perfumery vaselin can be used like fat or oil, over which it has the advantage in that it always remains odorless and free from acid; hence it is very appropriate for the manufacture of pomades. The market affords numerous varieties of this substance, under different names: vaselin (oil and solid), albolene (oil and solid), cosmolin, etc., etc.

SPERMACETI

is a substance found in the skull cavities of several whales and dolphins. In its properties it stands midway between beeswax, paraffin, and firm fats. In the living animal spermaceti is fluid, but after its death it congeals to a white crystalline mass of a fatty lustre, which melts at 40° C. (104° F.), and is frequently used for fine candles as well as for other articles.

WAX

(Cera alba), the well-known product of the bee; in perfumery only bleached (white) wax is employed. In recent years Japanese wax has appeared in commerce; this is of vegetable origin, but in its properties resembles beeswax.

SUBNITRATE OF BISMUTH,

bismuth white, pearl white, bismuthi subnitras, blanc de bismuth, blanc de perles, the basic nitrate of bismuth, the chief ingredient of many skin cosmetics, is prepared by dissolving metallic bismuth in moderately strong nitric acid, and pouring the solution into a large quantity of water, whereupon the subnitrate is precipitated.

The precipitated powder is collected on a funnel and washed with pure water until the wash water no longer changes blue tincture of litmus to red. The bismuth white is dried and preserved in well-closed vessels, since in the air it gradually assumes a yellowish color; for any sulphuretted hydrogen present in the air is greedily absorbed by this salt, and the resulting combination with sulphur has a black color.

OXIDE OF TIN

is obtained by treating metallic tin with fuming nitric acid, adding the solution to a large quantity of water, and washing the product, which forms a white insoluble powder used cosmetically for polishing the finger nails.

Besides the chemical products here enumerated, some others find application in perfumery; we shall describe their properties in connection with the articles into which they enter. In this connection mention may be made of the fact that more and more aromatic substances are now made artificially which were formerly obtained with difficulty from plants. Besides vanillin mentioned above, cumarin, oil of wintergreen, and some other products are prepared artificially. Heliotropin and nerolin are artificially prepared substances, possessing an odor resembling that of heliotrope and oil of neroli, respectively, but not identical chemically with the natural odorous substance. Artificial musk (Baur's), is playing a rôle at present, but is not identical with the natural substance.

C. The Colors used in Perfumery.

Some articles are colored intentionally; this remark applies particularly to some soaps which not rarely are stained to cɔrrespond to the color of the flower whose odor they bear; for instance, violet soap. Some articles again are used only on account of their color; for instance, paints, hair and whisker dyes. As we shall discuss this subject at greater length in connection with these toilet articles, we merely state here that nowadays every manufacturer can choose between a large number of dyes of any color, all of which are innoxious; hence no perfumer should under any circumstances use poisonous colors. This is a matter of importance with substances intended for immediate contact with the human body such as paints, lip salves, soaps, etc. All of these colors will be described hereafter.

CHAPTER VII.

THE EXTRACTION OF ODORS.

EXCEPTING the articles made in Turkey and India (especially oil of rose), most aromatic substances are manufactured in southern France and the adjoining regions of Italy, while a few (oils of peppermint and lavender) are produced in England; a few also (oils of peppermint, spearmint, wintergreen, sassafras, etc.) in the United States.

The methods by which the odors can be extracted from the plants differ somewhat according to the nature of the raw material and the relative delicacy or stability of the odorant principles. The older method was by distillation with steam and was prac-

ticed by the Arabian chemists of the fourth century. Later on
this process, which was found unsuitable for the extremely delicate odors of several flowers such as the tuberose and jasmine,
was supplemented by the slow and costly enfleurage method,
which extracted slowly the odor with pure cold lard. A still
later development, the full importance of which is not yet fully
appreciated, was the discovery of the volatile solvent process,
which made possible for the first time the production of the so-
called hyperessences and floressences, liquid products convenient
to use and absolutely faithful to the perfumes of the flowers
from which they are derived.

PRESSURE.

Certain aromatic substances that occur in large amounts
in some parts of plants, are best obtained by pressure. The
rinds of certain fruits contain an essential oil in considerable
quantities inclosed in receptacles easily distinguished under
the microscope. When these vegetable substances are subjected to strong pressure, the oil receptacles burst and the
essential oil escapes. The force is usually applied through a
screw press with a stout iron spindle; the vegetable substances being inclosed in strong linen or horse-hair cloths,
placed between iron plates, and subjected to a gradually increasing pressure. Comparative experiments have shown us
that even with the most powerful presses a considerable
amount of oil is lost owing to the fact that a large number of
oil receptacles remain intact. For this reason, when oil is to
be extracted by pressure, a hydraulic press is preferable, as
it develops greater power than any other press. In the hydraulic presses used for this purpose the piston fits exactly
into a hollow iron cylinder with sieve-like openings in its cir-
cumference. The vegetable substances are filled into this
cylinder; when the pressure is applied, the fluids escape

through the perforations, and the residue forms a compact woody cake which is then free from oil.

Besides the essential oil, watery fluid is expressed, the whole appearing as a milky liquid, owing to the admixture of vegetable fibres, mucilage, etc. It is collected in a tall glass cylinder which is set in a place free from any vibration. After remaining at rest for several hours the liquid separates into two layers, the lower being watery and mixed with

FIG. 4.

mucilage, that floating on top being almost pure oil. The latter is separated, and finally purified by filtration through a double paper cone in a funnel covered with a glass plate.

It is best to separate the water and oil in a regular separatory funnel, or in a simple apparatus illustrated in Fig. 4. It is made by cutting the bottom from a tall flask, and fitting into the neck by means of a cork a glass tube having a diameter of one-fourth to one-half inch. A rubber tube with stop-cock is fastened to the glass tube. By careful opening of the stop-cock, the watery fluid can be drained off to the last drop.

To the perfumer this method is of little importance, since it is applicable only to a few substances which, moreover, give cheap odors. Still, the possession of a hydraulic press is advisable to every manufacturer who works on a large scale, as it is useful also in the preparation of several fixed oils fre-

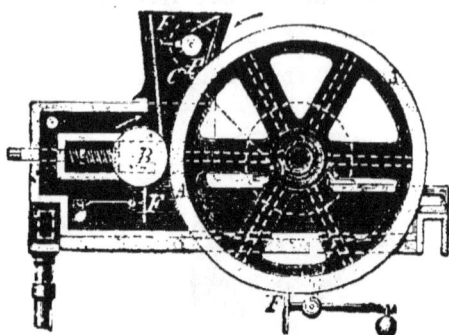

FIG. 5.

quently employed in perfumery, for instance, oils of almonds, nuts, etc.

Fixed oils are best extracted in so-called drop presses, the material having first been comminuted between rollers. These are arranged as shown in section in Fig. 5, and in

FIG. 6.

ground plan in Fig. 6. The apparatus consists of two smooth or slightly grooved iron cylinders A and B, respectively four feet and one foot in diameter, which can be approximated or separated by means of set screws. The material is placed into the trough F containing a feeding roller moved by the belt P. The scrapers FF, pressed against the cylinders by means of weighted levers, free the rollers from adhering pieces.

The drop presses Figs. 7 and 8 consist of a hydraulic press with cylinders A and piston B; the troughs E are movable by means of rings between two vertical columns and every trough has a circular gutter *d* for the reception of the expressed oil. The iron pots G have double walls, the inner of which has a series of openings at its upper part; these pots are filled with

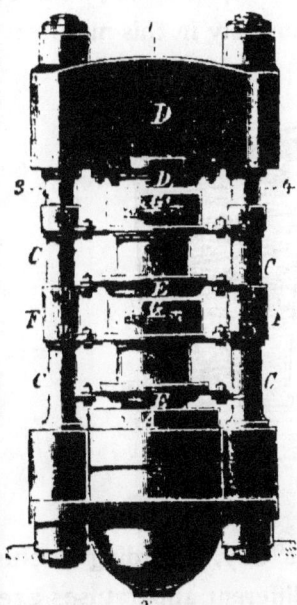

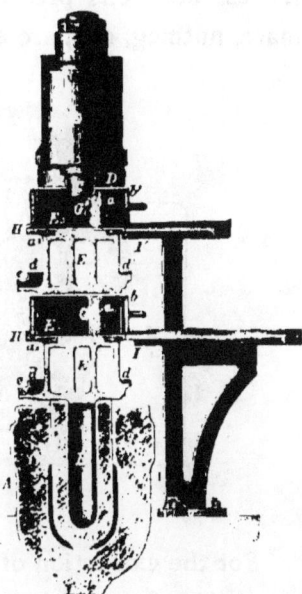

FIG. 7.   FIG. 8.

the bruised material to be pressed and after this has been covered with a plate of horse-hair tissue are set in the press.

As the piston rises, the troughs E sink into the pots, the escaping oil collects in the gutters *d* and thence passes into a receptacle. After pressing, the piston is allowed to sink back, the pots G are drawn aside (Fig. 8) to tabular surfaces, and other pots are substituted for the exhausted ones. These drop presses are suitable for the extraction of all fixed oils and also volatile oils present in orange and lemon peel, etc.

DISTILLATION.

Many odors or essential oils possess the remarkable property that their vapors pass so largely with that of boiling water that they can be extracted in this way (by "distillation") from vegetable substances, though the essential oils have a boiling-point far above that of water. Distillation can be employed for a large number of substances; for instance, the essential oils present in cumin, anise, lavender, fennel, mace, nutmeg, etc., are extracted exclusively in this manner.

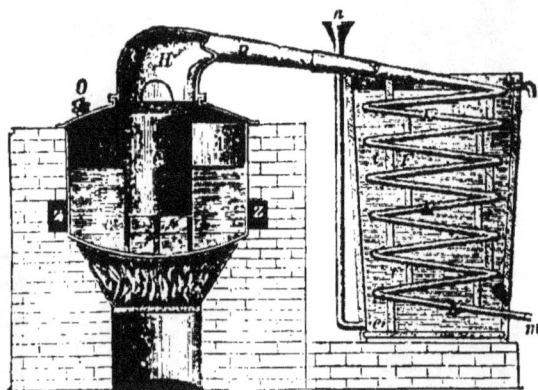

Fig. 9.

For the extraction of odors in this way, according to the quantities of material to be worked, different apparatuses are used, some of the most important of which will be here described.

For manufacturers who run without steam and are obliged to use a naked flame, the adjoining apparatus (Fig. 9) will be advantageous.

It consists of a copper boiler A, the still, set in a brick furnace. The latter is so constructed that the incandescent gases strike not only the curved bottom of the still, but also its sides through the flues Z left in the brickwork. The still, whose upper part projects from the furnace, has an opening O

on the left side, closed air-tight with a screw, which serves
for refilling with water during distillation when necessary.
To the margin of the still is fitted steam-tight the helm H,
made of copper or tinned iron, having a prolongation, the
tube R. The latter is joined to the conical projection v
which terminates in the worm K. In some apparatuses this
projection is omitted and the tube immediately joins the
worm. The latter is made of tinned iron and, as the cut
shows, is arranged in coils and supported by props t in the
wooden or metal condenser F. The condenser bears above a
short bent tube b, and below, immediately over the bottom,
an elbow tube e, long enough to reach above the edge of the
condenser, as indicated in the cut.

The vegetable substances to be distilled can be put immediately into the still and covered with water; but in this
case it is advisable to use a stirrer which must be. kept moving until the water boils, other-
wise the material might burn
at the bottom. But this accident can also be prevented by
applying a perforated false
bottom to the still above the
flues, or by inclosing the material in a wire-sieve basket C.

In place of the basket C
the apparatus can also be provided with an additional vessel containing the material to
be distilled. In the still A
(Fig. 10) the water is brought

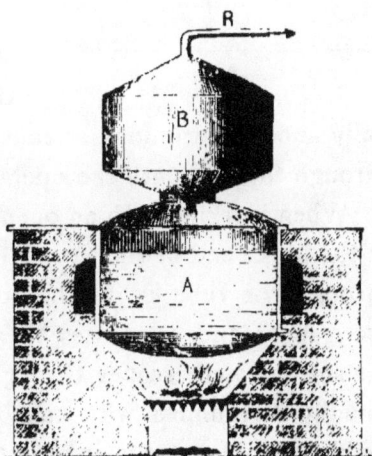

FIG. 10.

to boiling, the steam rises through the second still B in which
the material is spread on a perforated bottom. The steam
laden with the vapors of the essential oil passes through the
tube R into the condenser.

It is very advantageous, and in large establishments altogether indispensable, to use steam in the distillation of essential oils. Fig. 11 represents the arrangement of such an apparatus. The still B (which in this case may be made of
stout tinned iron) stands free and is provided with a wooden
jacket M for the purpose of retaining the heat. Immediately
above the curved bottom is a
perforated plate on which the
material rests. The tube D
which enters the bottom of the
still is connected with the boiler
which furnishes steam at moderate tension. H is the faucet
for the admission of steam; H,
is the faucet by which the water
escapes from the still at the
end of the operation. After the
still is filled with the material,
the faucet H is opened grad-
ually and a continuous stream of steam is allowed to pass
through the still until the operation is finished.

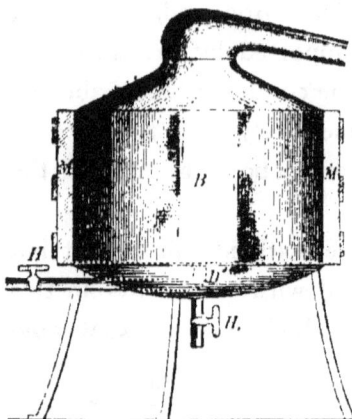

FIG. 11.

When working with an open fire, as soon as vapors appear
at the lower end of the worm (Fig. 9), cold water is admitted
through the tube *ne;* as the cold water abstracts heat from
the vapors and condenses them, it becomes warm, rises to
the surface, and escapes through *b*, so that the worm is continually surrounded with cold water. If for any reason the
saving of cold water is an object, its flow may be so regulated
that the vapors are just condensed, the warm distillate being
allowed to cool in the air. When working with steam, the
cold water must be admitted the moment the steam-cock is
opened, and the flow of cold water should be ample during
the distillation, which in this case is much shorter.

The large apparatuses here described are generally used,

especially for the extraction from vegetable substances of odors present in considerable quantity, for instance, mace, nutmeg, cloves, cinnamon, etc., or from bulky material as the various flowers. For very expensive odors, smaller apparatuses are often employed, the construction of which resembles that of the ones described. For this purpose small glass apparatuses are very suitable; they are illustrated in Fig. 12.

The still, a retort A, consists of a spherical vessel with a bottle neck *t* which is either closed with a cork or carries a thermometer or glass tube, and with a lateral tube, the neck

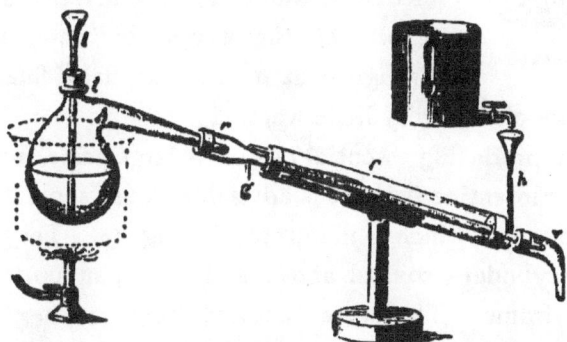

FIG. 12.

of the retort, connected with the adapter *r*. The latter passes into the condenser C. At the lower end of R is the bent adapter *v* under which is placed the receptacle for the distillate. The tube C is closed with corks, at its lower end is the ascending tube *h*, and at its upper end the descending tube *g*. During the distillation cold water flows in through *h* which cools the tube *r* and escapes at *g*. The tube C, as will be readily understood, acts like the condenser in the larger apparatuses above described. In order to prevent the breaking of the retort, it is not heated over a flame, but is set in a tin vessel B filled with water. The comminuted vegetable material is inserted with water through the up-turned neck of the retort into the latter; the vessel B is filled with water which is raised to the boiling-point.

During distillation we obtain at the lower end of the condenser pure water and essential oil. When larger quantities
are to be distilled it is advisable to use a Florentine flask as
a receptacle for the separation of the oil and water (Fig. 13).

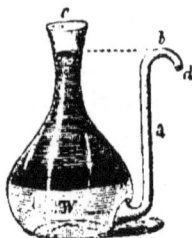

FIG. 13.

It consists of a glass bottle from the bottom
of which ascends a tube curved above; the
latter rises high enough to bring the curv-
ature slightly below the neck of the flask.
During the distillation the flask becomes
filled with water W, on which floats a layer
of oil O; the excess of water escapes
through *a* at *d* until the flask finally contains more oil and very little water.

When producing essential oils on a large scale, instead of
the frail Florentine flasks it is advisable to use separators, the
construction of which is illustrated in Fig. 14. They consist
of glass cylinders, conical above and below, supported on a
suitable frame. The water accumulating
under the oil is allowed to escape by opening the stop-cock; when the first separator
is filled with oil, the succeeding distillate
passes through the horizontal tube into the
next separator, etc.

When the distillation is carried on in an
ordinary still, we obtain, besides the essential oil, a considerable quantity of aromatic
water, that is, a solution of the oil in water.
An apparatus which obviates the losses
caused thereby is that of Schimmel described below, which is well adapted to

FIG. 14.

the manufacture on a large scale. The apparatus is patented.

The nearly spherical still D (Fig. 15) is surrounded by a
jacket M; the inlet steam tube R is connected with a branch
r which enters the interior of the still as a spiral tube with

numerous perforations, while R opens into the space M. When *r* is opened, distillation takes place by direct steam; when R is opened, by indirect steam; when both faucets are opened, the still is heated at the same time with direct and indirect steam.

The vapors rising from the still D pass through the helm C and the tube A into the worm K; the fluid condensed in

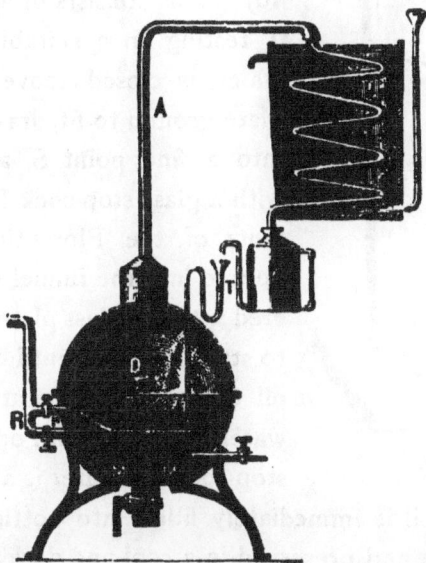

FIG. 15.

the latter drops into the tin Florentine flask F, the aromatic water flowing from the latter passes back into the still D through the Welter funnel T and is distilled over again, so that the entire distillation can be effected with very little water, and it is continued until the water escaping from the Florentine flask is freed from oil and odorless.

When working with superheated steam, it is necessary to set under the funnel tube T a vessel twice the size of the Florentine flask, which is provided with a stop-cock above and below. The lower cock is closed, the vessel is allowed to fill with water from F, then the upper cock is closed, the con-

7

tents being allowed to escape into D by opening, when the cocks are again reversed.

The use of superheated steam is important especially with material which gives up the contained oil with difficulty, such as woods.

For freeing the essential oil completely from water we use a so-called separating funnel (Fig. 16). This consists of a glass funnel T resting on a suitable support G, which is closed above with a glass plate ground to fit, drawn out below into a fine point S, and provided with a glass stop-cock H. The contents of the Florentine flask are poured into the funnel which is covered with the glass plate and allowed to stand at rest until the layer of oil O is clearly separated from the water W. By careful opening of the stop-cock the water is allowed to escape and the oil is immediately filled into bottles which are closed air tight and preserved in a cool and dark place.

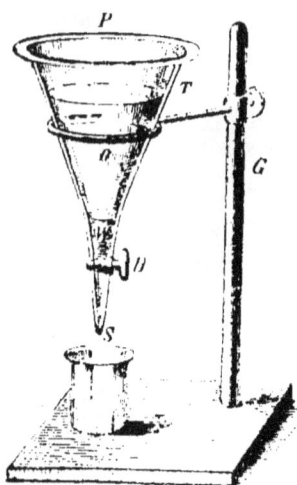

FIG. 16.

MACERATION (INFUSION).

Some odors, like those of cassie, rose, reseda, syringa, jasmine, violets, and many other fragrant blossoms, cannot be obtained by distillation as completely or as sweet-scented as by the process of maceration which is in general use among the large perfumers in southern France. This process is based on the property of fats to absorb odorous substances with avidity and to yield them almost entirely to strong alcohol. According to the fat employed for the maceration of the flowers—a solid fat like lard or a liquid like olive oil—odorous products are obtained which are known either as pomades or as perfumed oils (huiles antiques). By repeatedly treating fresh flowers with the same fat the manufacturer is able to perfume the pomade or oil at will, and in the factories these varying strengths are designated by numbers: the higher numbers indicating the stronger products.

The process of maceration is very simple. The fat is put into porcelain or enamelled iron pots which are heated, in a shallow vessel filled with water, to 40 or at most 50° C. (104 –122° F.); the flowers are inclosed in small bags of fine linen and hung into the fat, where they are allowed to remain for from one-half to two days. At the end of that time the bags are removed, drained, expressed, refilled with fresh flowers, and replaced in the fat. This procedure is repeated twelve

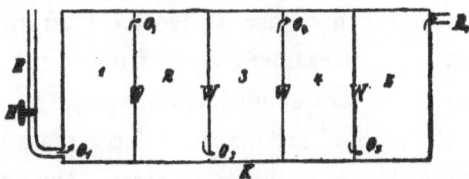

FIG. 17.

to sixteen times or oftener, thus producing pomades or oils of varying fragrance.

As the odors are much superior when the flowers are only a short time in contact with the fat, it is better to use an apparatus for continuous operation (Fig. 17). It consists of a box K made of tin plate, which is divided into from five to ten compartments by vertical septa and can be closed water tight by a lid to be screwed on. The septa have alternate upper and lower openings. The compartments contain each a basket of tinned wire filled with the flowers for maceration, then the lid is closed and the box heated in a water bath to 40 or 50° C. (104–122° F.). The stop-cock H in tube R is now opened. This admits melted fat or oil from a vessel above to the first compartment in which it rises through the basket filled with flowers whose odor it abstracts. The additional fat coming from above drives it over through the opening O_2 into compartment 2, where it comes in contact with fresh flowers, passes through O_3 into the third compartment, and so on through 4 and 5, until it finally escapes through R_1 well charged with odor. According to requirements a larger number of compartments may be employed.

When all the fat has passed through the apparatus, it is opened, the basket is removed from compartment 1, the basket from No. 2 is placed in 1, that from 3 in 2, etc.; basket 1 is emptied, filled with fresh flowers, and placed in compartment 5, so that every basket gradually passes through all compartments to No. 1. In this way the fat rapidly absorbs all the odor.

The odorous substances are abstracted from the pomades or huiles antiques by treatment with strong alcohol (90–95%) which dissolves the essential oils but not the fats. The huiles antiques with the alcohol are placed in large glass bottles and frequently shaken. In order to abstract the odors from pomades, the latter are allowed to congeal and are divided into small pieces which are inserted into the bottles of alcohol. A better plan is to fill the pomades into a tin cylinder with a narrow opening in front and to express the pomades, by a well-fitting piston, in the shape of a thin thread which thus presents a large surface to the action of the alcohol, thus hastening the absorption of the odor. The alcoholic solution obtained after some weeks is then distilled off at a low temperature. We shall recur to this hereafter.

No matter how long the fats are left in contact with alcohol, they do not yield up to it all the odor, but retain a small portion of it and hence have a very fragrant smell. They are, therefore, brought into commerce as perfumed oils or pomades bearing the name of the odorous substance they contain: orange flower, reseda pomade or oil, etc.; they are highly prized and are sometimes used again for the extraction of the same odor.

Some odors cannot bear even the slight rise of temperature necessary for their extraction by the method of maceration or infusion. For these delicate odors one of the following methods may be employed.

ABSORPTION OR ENFLEURAGE.

In this method the absorbing power of fat is likewise used for retaining the odors, but the flowers are treated with the fat at ordinary temperatures. This procedure which is employed especially in southern France is carried out as follows. The fat (lard) is spread to a thickness of about one-quarter

FIG. 18.

inch on glass plates G one yard long and two feet wide, which are inserted in wooden frames R and sprinkled with flowers F (Fig. 18). The frames are superimposed (the cut shows two of the frames) and left for from one to three days, when fresh flowers are substituted for the wilted ones, and so on until the pomade has attained the desired strength.

This procedure is very cumbrous and tedious and therefore had better be modified thus: In an air-tight box K (Fig. 19) we place a larger number of glass plates g covered with lard drawn into fine threads by means of a syringe. This box is connected with a smaller one K_1 which is filled with fresh flowers and provided with openings below and above, O and O_1. The latter, O_1, communicates by a tube with box K, at whose upper end is a tube e terminating in an exhaust fan so that the air must pass through the apparatus in the direction indicated by the arrows. A small fan V driven by clockwork will answer. The air drawn from K_1 is laden with odors and in passing over the fat as shown by the arrows gives them up completely to the fat. The use of this apparatus has very important advantages: the absorption is effected rapidly, requires little power, and the flowers do not come at all into

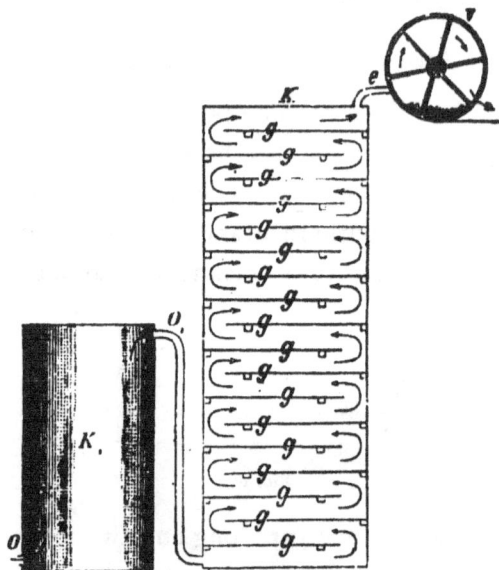

Fig. 19.

contact with the fat which therefore can take up nothing but the odors present in the air.

Instead of charging the fat with odors by either one of the methods here described, carbonic acid can also be employed with advantage, by means of the apparatus illustrated in Fig. 20. The large glass vessel G contains pieces of white marble M upon which hydrochloric acid is poured at intervals through the funnel tube R. A current of carbonic acid is thus developed, which passes through a wash bottle W filled with water, then through the tin vessel B containing fresh flowers, and finally into a bottle A filled with strong alcohol and set in cold water, after which it escapes through the tube

The carbonic acid absorbs the aromatic vapors from B and leaves them in the alcohol which absorbs them. (G, R, W are made of glass, B of tin.)

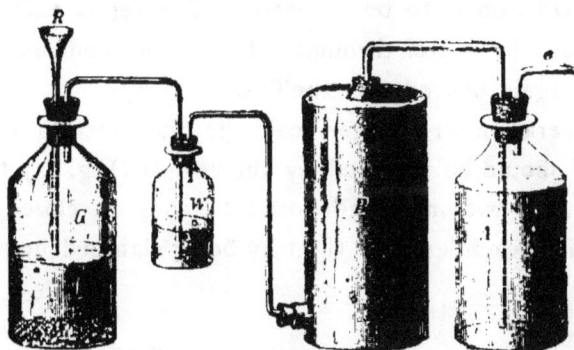

FIG. 20

EXTRACTION.

This method is based upon the discovery that carefully purified petroleum ether will rapidly extract the aromatic principles of flowers and plants. Evaporation of the extract under a gentle heat and a partial vacuum leaves the solid concretes which contain all the odorant principles of the raw material plus a certain amount of inert waxes. From these concretes the method of Dr. Charabot permits the isolation of the pure perfume oils, free from all inert materials and remarkable for their fidelity to the odor of the living flowers. This process, while simple in principle, is, in practice, one of the greatest delicacy and few have been able to approach the success of the originator.

The apparatus we use for this purpose is illustrated in Fig. 21. It consists of a Cylinder C made of tinned iron, which is provided above with a circular gutter R terminating in a stopcock h and which can be closed by a lid D bearing a stop-cock o. A tube b with a stop-cock a enters the bottom of the cylinder. The latter is filled with the flowers, the volatile liquid (petroleum ether, bisulphide of carbon, etc.) is poured over them, the lid is put on, and the gutter R filled with water, thereby sealing the contents of the vessel hermetically.

After the extraction, which requires about thirty to forty minutes, stop-cock *o* is opened first, then stop-cock *a*, and the liquid is allowed to escape into the retort of the still (Fig. 12). If the extraction is to be repeated, the water is allowed to escape from the gutter through *h*, the lid is opened, and the solvent is again poured over the flowers.

For operation on a larger scale the glass retorts are too small and should be replaced by tin vessels (Fig. 22) having the form of a wide-mouthed bottle F; they are closed by a lid D which is rendered air tight by being clamped upon the

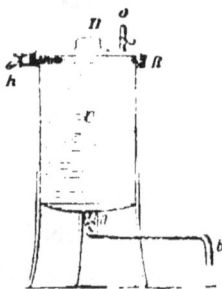

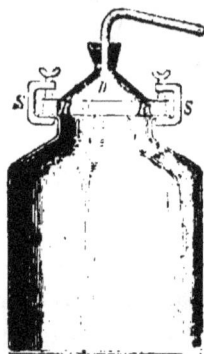

FIG. 21.   FIG. 22.

flange of the vessel (R) with iron screws S, a pasteboard washer being interposed; a curved glass tube connects the apparatus with the condenser of Fig. 12.

The solutions of the aromatic substances are evaporated in these apparatuses at the lowest possible temperature, the solvent being condensed and used over again. The heat required is for ether about 36° C. (97° F.), for choloroform about 65° C. (149° F.), for petroleum ether about 56° C. (133° F.), and for bisulphide of carbon about 45° C. (113° F.). If it is desired to obtain the aromatic substances pure from an alcoholic extract of the pomades made by one of the above-described processes—which is rarely done since these solutions are generally used as such for perfumes—a heat of 75 to 80° C. (167 to 176° F.) is required.

Another extraction apparatus illustrated in Fig. 23 is well adapted to operations on a large scale. Its main parts are the extractor E and the still B. The former is set in a vat W continually supplied with cold water. The still B is surrounded with hot water in the boiler K.

To start the apparatus the cone C is removed, the vessel E is filled with the material to be extracted, and C is replaced. The faucets H_2 and H_4 are opened, the solvent is poured into the still through the latter, when these faucets are closed and those marked H and H_1 are opened.

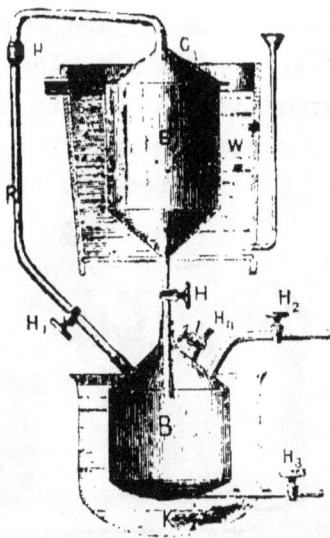

Fig. 23.

The water in K is heated until the contents of B are in brisk ebullition; the vapor rises through RH, is condensed on entering E and falls in small drops on the material. This fine rain of the solvent dissolves the aromatic substances and flows back into B, where it is again evaporated, and so on.

At the end of the extraction the faucets H and H_1 are closed and H_2 is opened. The vapors of the solvent pass through it into a worm where they are condensed; the essential oil remaining in B is drained off by opening H_3.

For still larger operations more perfect apparatuses are employed, such as those of Seiffert and Vohl. Seiffert's apparatus (Fig. 24) consists of a battery of jacketed cylinders; steam circulates in the space between the cylinders and the jackets. Each cylinder contains a plate covered with a wire net on which the flowers to be extracted are placed. All the cylinders having been filled and closed, the solvent is admitted from a container above, through S and a into C^2; when this is filled the liquid flows through $a^2b^3c^n$ into C . The solution saturated with essential oil leaves the apparatus through d^n and p and enters a reservoir. The course of the liquid is aided by the suction of an air-pump acting on p.

When the reservoir contains an amount of fluid equal to that in C^n, d^n is closed, a^n is opened, and C connected with C^1 through b^n and c^1. That the contents of C^2 are completely extracted is shown by the fact that the liquid appears colorless in the glass tube inserted in b^2; a^1 and C^2 are closed; a^2 and C^3 are opened, thereby excluding C^2 from the current of

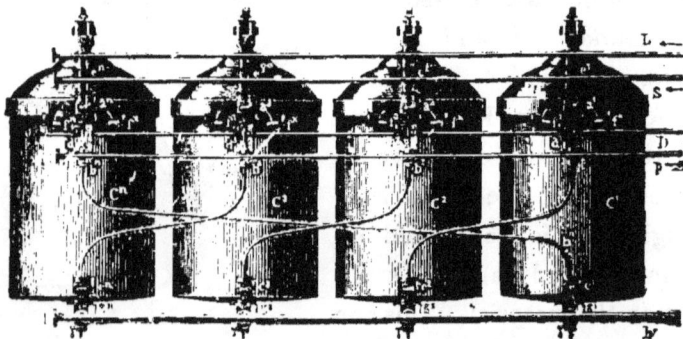

FIG. 24.

bisulphide of carbon which then flows through $C^3C^nC^1$. In order to permit the free flow of the bisulphide of carbon through S despite the exclusion of C^2, the faucets $a^1a^2a^3a^n$ must be two-way cocks; in one position they connect S with b; in the other they close b and leave the passage through S open.

In order to collect the bisulphide of carbon present in the extracted residue in C^2, faucet g^2 is opened and the bisulphide of carbon allowed to escape through h. The faucet e^2 in tube L on being opened admits compressed air to C^2, thus hastening the outflow. If nothing escapes below, faucets f^2 and f^x are opened, steam enters through tube D between jacket and cylinder; the bisulphide of carbon vapor passes through g^2

and *h* into the worm.   After the expulsion of the bisulphide
of carbon, C³ is emptied, refilled, connected with C¹, and bisulphide of carbon admitted from C³ in the manner above
described.

An extraction apparatus which has been much recommended of late is the so-called " Excelsior Apparatus " made
by Wegelin and Huebner, Halle a. S., which can be worked

FIG. 25.

with any desired solvent.   The construction of the apparatus
(Figs. 25 and 26) is as follows.

The solvent is admitted to the reservoir R in the lower
part of the condenser B through the tube indicated in the
figure.   The material to be extracted having been filled into
the cylinder A through the manhole, the apparatus is closed.
The cold water is admitted to the condenser by opening a
faucet.   The three-way cock shown in Fig. 25 is so placed as

to open a communication of the overflow tube with A. The faucet at the lower end of the reservoir R is now opened sufficiently and the solvent passes into A from above, and as it descends takes up more and more oil, flows through the sieve-

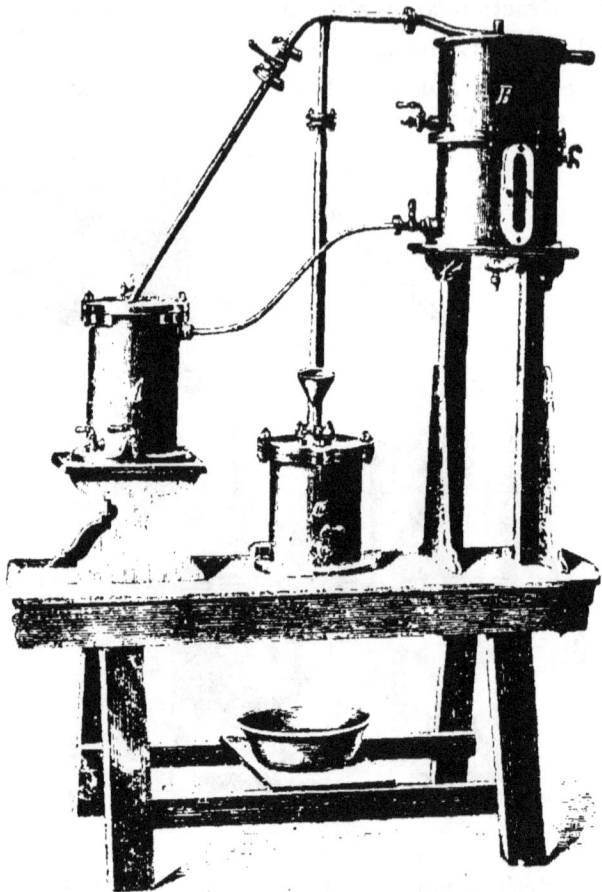

FIG. 26.

plate, and escapes through the tube at the bottom of A through the three-way cock, the overflow tube, and the drain tube into the accumulator C. The opening of a faucet now admits steam to the heating coil, when the solvent evaporates, leaving the oil or fat behind. It is condensed in B, again re-

turns to R, whence it passes once more through the faucet
into the extractor A. The vessel C and the tubes leading to
A and C are surrounded with felt to prevent loss of heat. A
sample taken from the small cock at the foot of A (it has a
small plate in the interior of the tube) will show when the extraction in A may be looked upon as finished. The solvent
is distilled off or recovered from the residue in A in the following manner. First the faucet in R is closed. The three-
way cock A is set to establish direct communication between

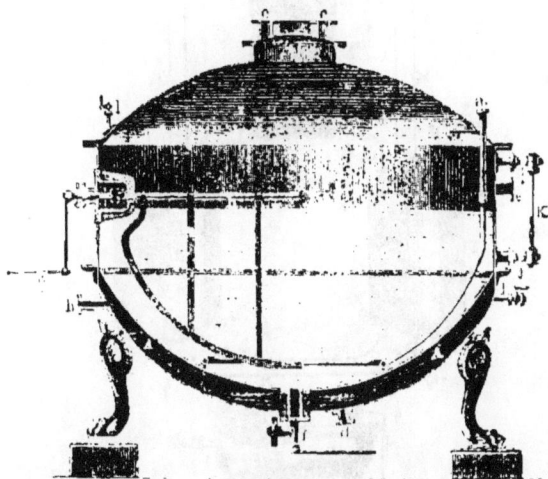

FIG. 27.

A and C, thus cutting off the overflow tube. Hence all the
solvent in A flows into C for distillation, while the oil is left
behind. Steam being admitted to the residue, the solvent
rises as vapor through the upper tube from A to B and collects in a liquid state in R. To drive off the last traces of
the solvent from the fat or oil obtained, steam is blown into
C by opening the valve. Besides the solvent, watery vapor
enters B and forms a layer of water in R under the solvent.
By taking a sample from the test-cock of the reservoir C
which has an internal small plate, the termination of the process is ascertained. The gauge tube at the reservoir shows

the level of the solvent and water. The water is drawn off
by opening the faucet at the lower end of the reservoir. A
is emptied through the manhole and by draining the oil from
C through the discharge cock. The tube R is closed by a
light valve so as to prevent evaporation of the solvent. All
the apparatuses work without pressure so that there is no
danger from overstrain.

The solutions of the essential oils in bisulphide of carbon

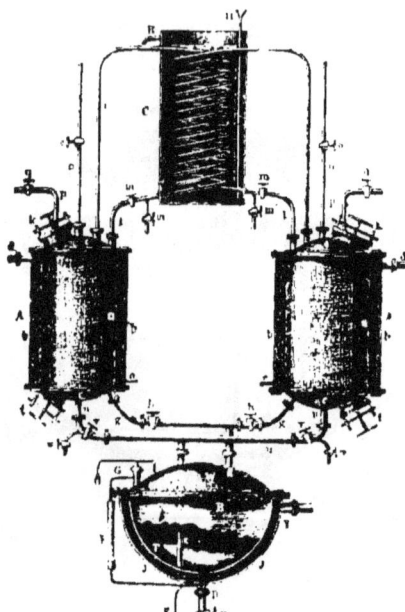

Fig. 28.

are distilled off in the steam still illustrated in Fig. 27; the
steam enters at *h*, the water of condensation escapes at *d*, the
liquid to be distilled enters at *e* from a container at a higher
level. The boiling is kept uniform by the stirring arrangement
hg. After the bisulphide of carbon is distilled off, air is passed
through the oil by the curved tube *a* which has fine perforations, so as to evaporate the last traces of the solvent.

In Vohl's apparatus (Fig. 28), arranged for petroleum
ether, the extraction is effected with the boiling fluid; hence

this apparatus is better adapted for the cheaper oils than for the finest oils from flowers. The apparatus consists of two extractors A A, the accumulator B, and the condenser C. Petroleum ether is allowed to flow over the substances to be extracted, by opening the faucets *mm*, *vh*, closing *ogw*E, and opening *o*, the course being through *ux* to B. When B is two-thirds full, the flow of petroleum ether is cut off, steam is admitted through *y* and the contents of B are brought to the boiling-point. The vapors pass through *g* and are condensed in *f* until the contents of A reach the boiling-point of the solvent, when the vapors pass through *i* into C, and after closing *m'* the liquid passes through *ml* into the inner cylinder of the extraction apparatus and returns through *uxx*.

After the contents of A are extracted, *m'* is opened, *m* closed, and steam is admitted through *d* into the jacket of A: the vapors of the solvent force the liquid part of the contents through *ux* into B. Overfilling of B is prevented by allowing the vapors of the solvent to escape at the proper time into the condenser through *p* by opening *q*. Then *v* is closed, *q* opened, and the steam present in A drawn off by an exhaust applied to *p*; as soon as *p* begins to cool, all the petroleum ether is distilled off, the steam is cut off at *d*, and the extract evacuated through *t*. The contents of B are brought into a still through D and E.

By employing greater pressure the extraction can also be effected by what is called displacement; the material to be extracted is placed in a stout-walled vessel S (Fig. 29) which is connected by a narrow tube at least ten yards long with the vessel F containing the solvent. Stopcock H is first opened, then stop-

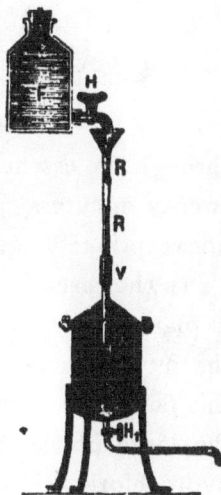

FIG. 29.

cock H, which is closed as soon as fluid begins to flow from it. After the liquid has remained in contact with the material for from thirty to sixty minutes, H, is opened very slowly, the liquid is allowed to escape and is displaced with water which is made to pass out of F in the same way as the solvent, until the latter is completely displaced from S.

After the solvent has been distilled off, the less volatile essential oil remains in the still almost pure, containing only traces of wax, vegetable fat or coloring matter which are of no consequence for our purposes. The last remnants of the solvent cannot be expelled by distillation, but by forcing

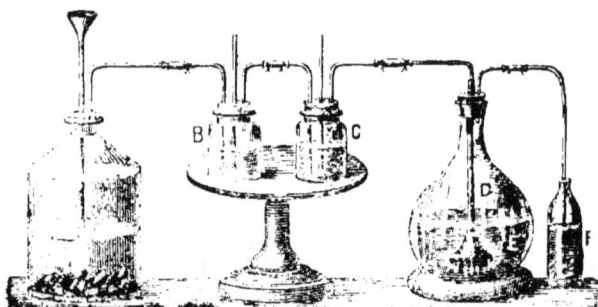

FIG. 30.

through the essential oil a current of pure air for fifteen or twenty minutes. The essential oils then are of the purest, unexceptionable quality.

In the case of delicate oils it is better to use carbonic acid in place of air for expelling the last traces of the solvent, as the oxygen may impair the delicacy of the fragrance. For this purpose we use the apparatus illustrated in Fig. 30. In the large bottle A carbonic acid is generated by pouring hydrochloric acid over fragments of white marble. The carbonic acid passes into the vessel B filled with water which frees it from any adhering drops of hydrochloric acid; then into C filled with sulphuric acid to which it yields its water so that only pure carbonic acid escapes through the fine rose at the end of tube D which is made of pure tin, and as it passes through the oil in E it carries off the last traces of the volatile solvent. In its final passage through the water in F it leaves behind any oil that may have been carried with it.

As all the aromatic substances change in air by the gradual absorption of oxygen, and lose their odor—become resinified—these costly substances must be put into small bottles which they completely fill, and be preserved in a cool dark place, as light and heat favor resinification. The bottles must be closed with well-fitting glass stoppers.

Aromatic waters or eaux aromatisées, such as jasmine water (eau de jasmin), orange-flower water (eau de fleurs d'oranges, eau triple de Néroli, aqua naphæ triplex), etc., are made by distillation of these flowers with water and show a faint but very fine odor. When they contain, besides, dilute alcohol they are called spirituous waters or esprits. Those brought into commerce from southern France are of excellent quality.

The Yield of Essential Oils.

The quantities of essential oil obtainable from the vegetable substances vary with the amount present in each. The following table shows the average quantities of oil to be obtained from 100 parts of material.

Material.	Name of Plant.	Mean Yield per 100 Parts.
Ajowan seed	Ptychotis Ajowan	3.000
Allspice	Myrtus Pimenta	3.500
Almonds, bitter	Amygdala amara	0.400–0.700
Angelica seed	Archangelica officinalis	1.150
Angelica root, Thuring	" "	0.750
" " Saxon	" "	1.000
Anise seed, Russian	Pimpinella Anisum	2.800
" " Thuring	" "	2.400
" " Morav	" "	2.600

8

Material.	Name of Plant.	Mean Yield per 100 Parts.
Anise seed, Chili........	Pimpinella Anisum	2.400
" " Spanish.....	" "	3.000
" " Levant	" "	1.300
Anise chaff...........	" "	0.666
Arnica flowers.........	Arnica montana.............	0.040
Arnica root...........	" "	1.100
Basilicum herb, fresh...	Ocymum basilicum	0.040
Bay leaves...........	Pimenta acris	2.300–2.600
Bergamots.......................		ab. 3.400
Betel leaves..........	Piper Betle.................	0.550
Bitter almond meal.....	Amygdala amara.............	0.950
Buchu leaves	Barosma crenulata............	2.600
Calamus root.........	Acorus Calamus.............	2.800
Camomile, German.....	Matricaria Chamomilla....	4.000–6.000
" Roman......	Anthemis nobilis	3.000
Caraway seed,		
Cult. German......	Carum Carvi.................	4.000
" Dutch	" "	5.500
" East Prussian.	" "	5.000
" Moravian.....	" "	5.000
Wild German......	" "	6.000–7.000
" Norwegian ...	" "	6.000–6.500
" Russian......	" "	3.000
Cardamoms, Ceylon.....	Elettaria Cardamomum	4.250
" Madras	" "	4.300
" Malabar....	" "	1.750
" Siam.......	" "	1.350
Cascarilla bark........	Croton Eluteria............	1.500
Cassia flowers.........	Cinnamomum Cassia..........	3.500
Cassia wood..........	" "	0.285
Cedar wood..........	Juniperus virginianus......	0.700–1.000
Celery herb..........	Apium graveolens............	0.200
Celery seed..........	" "	0.300

Material.	Name of Plant.	Mean Yield per 100 Parts.
Cinnamon, Ceylon......	Cinnamomum zeylanicum..	0.900–1.250
" white.......	Canella alba................	1.000
Cloves, Amboina.......	Caryophyllus aromaticus........	19.000
" Bourbon ..·.....	" "	18.000
" Zanzibar.......	" "	17.500
" stems.........	" "	6.000
Common wormwood herb.	Artemisia Abrotanum..........	0.040
" " root.	" "	0.100
Copaiva balsam, Para ...	Copaifera officinalis............	45.000
" " East Ind.	Dipterocarpus turbinatus.......	65.000
Coriander seed,		
Thuringian.......	Coriandrum sativum...........	0.800
Russian...........	" "	0.900
Dutch	" "	0.600
East Indian.......	" "	0.150
Italian...........	" "	0.700
Mogadore........	" "	0.600
Crisp mint herb.......	Mentha crispa................	1.000
Cubebs..............	Piper Cubeba..........	12.000–16.000
Cumin seed, Mogadore..	Cuminum Cyminum...........	3.000
" " Maltese....	" "	3.900
" " Syrian.....	" "	4.200
" " East Indian	" "	2.250
Curcuma root....:......	Curcuma longa...............	5.200
Dill seed, German......	Anethum graveolens..........	3.800
" " Russian......	" "	4.000
" " East Indian...	Anethum Sowa...............	2.000
Elder flowers	Sambucus niger	0.025
Elemi resin...........	Icica Abilo	17.000
Eucalyptus leaves, dry..	Eucalyptus globulus	3.000
Fennel seed,		
Saxon.............	Foeniculum vulgare.......	5.000–5.600
Galician..........	" "	6.000
East Indian.......	Foeniculum Panmorium........	2.200
Galanga root	Alpinia Galanga..............	0.750
Galbanum resin.......	Galbanum officinale..........	6.500
Geranium	Pelargonium odoratissimum.....	0.115

Material.	Name of Plant.	Mean Yield per 100 Parts.
Ginger root,		
African	Zingiber officinale	2.600
Bengal	" "	2.000
Japan	" "	1.800
Cochin China	" "	1.900
Hop flowers	Humulus Lupulus	0.700
Hop meal, lupulin	" "	2.250
Hyssop herb	Hyssopa officinalis	0.400
Juniper berries,		
German	Juniperus communis	0.500–0.700
Italian	" "	1.100–1.200
Hungarian	" "	1.000–1.100
Laurel berries	Laurus nobilis	1.000
Laurel leaves	" "	2.400
Laurel, Californian	Oreodaphne californica	7.600
Lavender flowers	Lavandula vera	2.900
Linaloe wood	Elaphrium graveolens	5.000
Lovage root	Levisticum officinale	0.600
Mace	Myristica fragrans	11.000–16.000
Marjoram herb, fresh	Origanum Majorana	0.350
" " dry	" "	0.900
Melissa herb	Melissa officinalis	0.100
Musk seed	Hibiscus Abelmoschus	0.200
Mustard seed,		
Dutch	Sinapis nigra	0.850
German	" "	0.750
East Indian	" "	0.590
Pugliese	" "	0.750

Material.	Name of Plant.	Mean Yield per 100 Parts.
Mustard seed, Russian	Sinapis juncea	0.500
Myrrh	Balsamodendron Myrrha	2.500–6.500
Myrtle	Myrtus communis	0.275
Nutmegs	Myristica fragrans	8.000–10.000
Olibanum resin	Boswellia, var. spec.	6.300
Opoponax resin	Pastinaca Opoponax	6.500
Orange peel, sweet	Citrus Aurantium	2.500
Orris root	Iris florentina	0.200
Parsley herb	Apium Petroselinum	0.300
Parsley seed	" "	3.000
Parsnip seed	Pastinaca sativa	2.400
Patchouly herb	Pogostemon Patchouly	1.500–4.000
Peach kernels	Amygdalus persica	0.800–1.000
Pepper, black	Piper nigrum	2.200
Peppermint, fresh	Mentha piperita	0.300
Peppermint, dry	" "	1.000–1.250
Peru balsam	Toluifera Pereiræ	0.400
Rhodium wood	Convolvulus Scoparius	0.050
Rose flowers, fresh	Rosa centifolia	0.050
Rosemary	Rosmarinus officinalis	1.550
Rue herb	Ruta graveolens	0.180
Sage herb, German	Salvia officinalis	1.400
" " Italian	" "	1.700
Santal wood,		
East Indian	Santalum album	4.500
Macassar	" "	2.500
West Indian	Unknown	2.700
Sassafras wood	Sassafras officinalis	2.600
Savin herb	Juniperus Sabina	3.750
Snakeroot, Canadian	Asarum canadense	2.800–3.250
" Virginian	Aristolochia Serpentaria	2.000
Star-anise, Chinese	Illicium anisatum	5.000
" Japanese	Illicium religiosum	1.000
Storax	Liquidambar orientalis	1.000
Sumbul root	Ferula Sumbul	0.300

Material.	Name of Plant.	Mean Yield per 100 Parts.
Tansy herb............	Tanacetum vulgare............	0.150
Thyme	Thymus Serpyllum.............	0.200
" dry	" "	0.100
Vetiver root..........	Andropogon muricatus.....	0.200–0.350
Violet flowers.........	Viola odorata	0.030
Wintersweet marjoram..	Origanum creticum............	3.500
Worm seed...........	Artemisia maritima	2.000
Wormwood herb	Artemisia Absinthium 0.300–0.400

Fresh flowers as a rule contain more aromatic material than wilted ones; the yield of dried herbs, leaves, etc., is usually greater than that of the fresh, because the latter contain much water which is lost in drying. When such vegetable materials cannot be worked fresh, which is best, they should be completely dried, spread on boards, at a moderate temperature in the shade and preserved in dry airy rooms, special care being had to guard against mould.

CHAPTER VIII.

THE SPECIAL CHARACTERISTICS OF ARO-MATIC SUBSTANCES.

In a preceding chapter on the origin of the many aromatic substances, many of their special characteristics have been lightly touched upon. In this one an attempt will be made to carry this essential study a little further, indicating the physical and other properties which serve to differentiate the perfume materials and by which their relative purity and value may be determined, or at least indicated.   This knowledge is of the utmost importance, since few if any of the essential oils are prepared by the user, who must purchase them in the open market, depending on his own experience and on the integrity of the firms with which he deals for the assurance that his materials are of satisfactory quality.   Adulteration is frequent and is almost universal in the cheaper grades and is frequently difficult of detection, and furnishes a subject which will be discussed at greater length elsewhere.

OIL OF CASSIE.

The oil derived from the flowers of Acacia farnesiana is seldom if ever found in commerce, this valuable odorant material being made by the pomade and extraction processes.   The pomade is still used largely but is partially superseded by the liquid flower essences prepared by the newer method of Charabot. There is no chemical criterion of purity and dependence must be placed on the nose test or, better, on the reputation of the producer.

OIL OF ANISE

should be colorless or faintly yellow; a dark yellow color indicates old and inferior quality.   The characteristics of this oil are the odor, its aromatic sweet taste, and especially the property of solidifying at a comparatively high temperature, 10-15° C. (50-59° F.), which is due to the separation of a stearopten, anethol.   Oil of anise is frequently adulterated with or replaced by oil of star-anise.   The easy solidification of the oil of anise is not always proof of its good quality, for the oil from anise chaff, which congeals at a still higher temperature, is sometimes mixed with it, and this has a less fine odor than that distilled from the seed.   One part by weight of oil of anise is soluble in an equal weight of alcohol of 94%.

OIL OF BERGAMOT

has a pale yellow color which becomes greenish when the oil is kept in copper vessels, and a strong agreeable odor.   This oil requires the greatest care in its preservation, as it abstracts oxygen from the air with extreme rapidity, when it changes its superior odor so that it can hardly be distinguished from oil of turpentine.

OIL OF BITTER ALMOND (OLEUM AMYGDALE AMARÆ),

when pure, is a colorless, refractive liquid which is heavier than water. The vessels in which this product is preserved must be stoppered air-tight, for in the air the oil very quickly changes into a white, odorless mass of crystals consisting of benzoic acid.

Oil of bitter almond is formed by the action of the amygdalin upon the emulsin present in the fruit, bitter-almond meal being deprived of fat and left in contact with water for some hours at from 40–45° C. (104–113° F.). Besides oil of bitter almond, sugar and prussic acid are likewise formed. The crude oil distilled from the meal is freed from the prussic acid by agitation with ferrous chloride and lime-water, and redistillation.

OIL OF CAJEPUT (OLEUM CAJUPUTI)

has usually a greenish color, and has a burning, camphoraceous and at the same time cooling taste. It has a peculiar odor resembling that of camphor and rosemary.

OIL OF CALAMUS (OLEUM CALAMI).

This oil, which is very viscid and of a yellow or reddish color, must usually be mixed with other essential oils in order to furnish pleasant perfumes.

OIL OF CHAMOMILE (OLEUM CHAMOMILLÆ).

Oil of chamomile, from Matricaria Chamomilla (common chamomile), which is specially characterized by its magnificent dark-blue color, has a marked narcotic odor and is very high-priced, owing to the small yield of oil by the flowers. The oil from Anthemis nobilis (Roman chamomile) has also a blue color which gradually becomes greenish-yellow.

CAMPHOR (CAMPHORA).

This essential oil differs from the others mainly by being firm and crystalline at ordinary temperatures. Chinese or Japanese camphor melts at 175° C. (347° F.) and boils at 205° C. (401° F.). Camphor is seldom used alone, as its odor is hardly fragrant; but it finds frequent application in the preparation of mouth washes, toilet vinegars, etc. In commerce so-called Borneo camphor is also met with (though rarely), which closely resembles the Chinese in appearance and other qualities, but is more friable and melts at 189° C. (388.4° F.).

OIL OF CASCARILLA

is not used pure in perfumery, the bark being generally employed instead.

OIL OF CASSIA (OLEUM CASSIÆ)

has a yellow color, gradually becoming dark reddish-brown, and an odor resembling that of oil of cinnamon, but the odor is not so fine, nor so strong, as that of the latter. The taste of the oil is of special importance: while that of true oil of cinnamon is burning though sweet, oil of cassia has a sharper taste, and this taste is considered by some a good mark of recognition of the rather common adulteration of true oil of cinnamon which is much more costly.

OIL OF CEDAR.

This oil, obtained from the wood of the Juniperus virginiana (not from the true cedar, Cedrus Libani), is clear like water, has a pleasant odor, and differs from most essential oils by congealing at a very low temperature (— 22° C. or — 8° F.), and by its uncommon resinification in contact with air.

OIL OF LEMON

is of importance to the perfumer as well as to the manufacturer of candy, flavoring extracts and beverages. While it is rarely employed to advantage in extracts or toilet waters and never in powders, it has a wide field of application in the scenting of the now extremely popular lemon soaps and creams, the odor being strengthened by the addition of specially purified citral. Lemon oil is pale yellow and of a strong, refreshing odor, which is soon lost in contact with air, when it acquires a disagreeable turpentine-like smell. This change is particularly marked under the influence of light. Its specific gravity is 0.850 at 20° C. The normal citral content is 3.5 to 5.5%, the remainder of the oil being composed mainly of terpenes and sesqui-terpenes, which have little odor. Owing to the presence of these terpenes the solubility in alcohol of normal strength is only 5%.

For use in soaps and creams only the best lemon oil obtainable will be found satisfactory, as the inferior oils develop a rancid odor in the product. The finest hand-pressed oil, as fresh as can be obtained, should be used. The distilled oil is suitable for flavoring purposes though not as desirable as the better grades. An unusually high citral content usually indicates fortification.

OIL OF LEMON-GRASS.

This oil, which, as the name indicates, has a lemon-like odor, is only useful in low-grade perfumes and in the cheap soaps, is imported chiefly from Ceylon. Its main importance to the perfume industry is due to its high citral content, the citral being separated and used as such or in making the synthetic ionones and other products. The citral content is usually from 65 to 85% and the oil is little subject to adulteration, but is occasionally employed to adulterate both lemon oil and geranium oil, oil of citronella being likewise employed for the latter purpose,

Oil of Coriander (Oleum Coriandri)

has a pale yellow color and a burning, sharp, aromatic taste. Like oil of cubebs (oleum cubebæ), oil of dill (oleum anethi), and oil of fennel (oleum fœniculi), which latter also has a rather low congealing point (— 8° C. or + 17° F.), this oil is used less in perfumery than for scenting soap and in the manufacture of liqueurs. But it should be noted that these oils, as well as those of bergamot, caraway, star-anise, and some others, could well be employed for cheap perfumes and for scenting soap. Oil of dill also finds application alone in the preparation of some face washes, and the dried fennel herb in cheap sachets.

Oil of Citronella

comes from Java and Ceylon chiefly, the former being of better quality. Like lemon-grass oil it is used only in cheap perfumes and soaps. It contains about 55% of total geraniol esters, i. e., geraniol, citronellol and their esters, and is important as a source of these compounds, which are much used in artificial rose oils. The chief adulterant which must be watched for is petroleum, and to detect it the Schimmel solubility test is used, which is based on the fact that the pure oil is soluble in 10 volumes of 80% alcohol.

Oil of Geranium.

As was mentioned earlier, it is necessary to distinguish between true Oil Geranium and Palmarosa Oil or Turkish Geranium, which is distilled from geranium grass or, in India, from ginger grass. Of the true geranium oil there are several varieties, the African or Algerian, which is considered the most desirable, Geranium Bourbon, obtained from the Bourbon Islands, and French Oil Geranium, which is less important on account of the smaller amount produced.

True Oil Geranium has a distinct rose odor to which it owes the title of rose geranium, and it contains considerable amounts of geraniol, one of the alcohols present in rose oils. It is accordingly used sometimes as a substitute for rose oil in cheaper perfumes and is important to the perfumer. The greatest care should be exercised to obtain the best grade and as adulteration is frequent and difficult to detect the reputation of the supply house is the best safeguard.

OIL OF ASPIC (OIL OF SPIKE LAVENDER).

Oil of Aspic, or Spike Lavender, is distilled from the flowers of Lavandula spica, a variety of lavender closely related to the Lavandula vera but producing an oil of widely varying character from true Oil Lavender and one of much lower value. Lavandula spica flourishes in the mountainous regions of Spain, France and Italy and the oil is distilled in those localities, the major portion coming from Spain. It is a pale yellow oil, with an odor resembling those of true lavender and rosemary, but unlike true Oil Lavender is practically free from esters and may be easily recognized by this fact. It contains, however, considerable percentages of linalol, camphor and borneol.

It is occasionally adulterated with turpentine, rosemary oil or camphor oil, and is itself used to adulterate true lavender oil, which is approximately four times as costly. Its chief use is in cheap perfumes and particularly in perfuming soaps, where it is used in place of all or part of the Oil Lavender, which would otherwise be employed. Adulteration of spike lavender is detected by chemical analyses, which are needed to supplement the simpler solubility test, the latter being only useful to detect adulteration by turpentine.

OIL OF JASMINE,

not to be confounded with the oil of Syringa or German jasmine (Philadelphus coronarius), is colorless or yellowish and has a very strong, almost narcotic odor. It is one of the most valuable and at the same time most expensive aromatic substances employed in perfumery. Genuine oil of jasmine can be obtained only at high prices, which are, however, justified not only by the expense of production but by the value to the perfumer of high grade oil of jasmine.

OIL OF CHERRY-LAUREL

is not used as such in perfumery; at most cherry-laurel water may be employed. But as this has the odor of oil of bitter almond and as the presence of some prussic acid, on account of which the officinal cherry-laurel water is used, is of no value to the perfumer and is, in fact, undesirable, owing to its poisonous quality, we substitute in all cases a corresponding quantity of oil of bitter almond for cherry-laurel water.

OIL OF CULILABAN (OLEUM CULILAVANI)

is light brown, somewhat viscid; the odor recalls that of the oils of cinnamon, sassafras, and clove. It has been used for scenting soap.

OIL OF CARAWAY (OLEUM CARI)

is light yellow and has an aromatic odor and burning taste. In perfumery it is used only for very cheap odors and for scenting soap; it finds its chief application in the manufacture of liqueurs.

OIL OF LAVENDER (OLEUM LAVANDULÆ).

This oil is obtained by the distillation of the flowers of Lavandula vera, growing principally in France, Italy and England. English Oil Lavender is more highly esteemed than the French varieties but is sold in much less volume, most of the commercial oil coming from French sources. The two varieties of oil are of quite different composition, the French containing from 25 to 44% esters, chiefly linalyl acetate, and the English only 7 to 10%. Other tests must, however, be used to distinguish the two, as the French oil is sometimes adulterated with spike lavender to bring the ester content down to that of the more expensive English oil. This form of adulteration fortunately affects the optical rotation as well as the ester content.

The French oil is also adulterated with spike lavender and sold as pure French Lavender, but the adulteration is easy of detection, as already mentioned. A more dangerous sort of sophistication is by stretching the true Oil Lavender Fleurs with artificial esters such as ethyl citrate, ethyl phthalate or glyceryl acetate, the esters being used in conjunction with spike lavender to stretch the oil and still maintain the proper ester content. Synthetic linalyl acetate is likewise used at times. Expertly sophisticated Oil Lavender is extremely difficult for the average perfumer to detect on account of the limited resources at his command for making extensive chemical tests, and, as in the case of so many other perfume materials, the best safeguard is to purchase oil the origin of which guarantees its authenticity.

OIL OF ROSEMARY.

Rosemary oil is distilled from the flower tops of the rosemary plant, Rosemarinus officinalis. The two principal varieties are the Italian and Dalmatian oils, which are distilled after the flower season is over, and the French oil, which is distilled from fresh flower tops. Rosemary of a very fine quality is also distilled in England from cultivated plants, while a lower grade is distilled in Spain from wild plants. The Spanish production is becoming of increasing importance and the quality is satisfactory if the oil is obtained from reliable sources. Adulteration of Oil Rosemary is not as frequent as of many other oils which are more expensive.

OIL OF ORRIS.

This oil is obtained by the distillation of orris root and on account of its odor, which closely resembles that of violets, is of the greatest importance to perfumers. In recent years, however, it has been found more satisfactory to use other orris preparations, such as resinarome orris and similar preparations, which preserve the delicate orris odor and which are valuable fixatives in violet compositions.

OIL OF SANDALWOOD.

The distillation of the tropical wood known as santal yields an oil of an aromatic, spicy odor which has a wide use in perfumes, particularly in the scenting of soap rather than in extracts and toilet waters. There are two important varieties of this oil, the East Indian and the West Indian, the former being the more expensive and of a better quality. Sandalwood oil distilled in England has the highest reputation for quality and freedom from adulteration, but the larger proportion is probably distilled nearer its origin.

OIL OF SWEET BAY (LAUREL) (OLEUM LAURI)

is green and usually mixed with the fixed oil of the same plant. It finds more frequent application in the manufacture of liqueurs than in perfumery; but as it has a pleasant odor it might well be used for cheap perfumes. But in that event it must be freed from the fixed oil by distillation.

OIL OF MAGNOLIA,

likewise, has not yet been prepared as such. The remarks made above under the head of oils of lily and wallflower apply also to this odor. The so-called magnolia perfumes are mixtures of different odors.

OIL OF MARJORAM (OLEUM MAJORANÆ)

Oil of marjoram, which is obtained by distillation from the dried herb, has a strong aromatic odor. It is mentioned as having often been used in perfumery for scenting soap instead of oil of thyme, whose odor, moreover, is very similar to that of marjoram, but this is a mistake, due to the fact that ordinary oil of thyme has long been sold under the name of oil of origanum. True oil of marjoram costs about twelve dollars a pound, while oil of tyme (so-called oil of origanum) is worth only about eighty cents. It is rarely employed for volatile perfumes.

OIL OF MELISSA.

The oil of Melissa officinalis, owing to the very small yield, is quite expensive. It is used only for the preparation of some perfumes which owe their peculiar qualities to this strong odor. This oil must not be confounded with the spurious oil of melissa, also called oil of citron-melissa, which is identical with oil of lemon grass (see page 30).

OILS OF MINT.

Although all the mints possess an agreeable odor, only three varieties find extensive application. There are the oils from Mentha piperita, peppermint; Mentha viridis, spearmint; and Mentha crispa, crispmint. The oils of English manufacture are highly esteemed, but the United States also produces them of excellent quality. At one time the cultivation of mints, particularly peppermint, was greatly extended, with the expectation of deriving satisfactory profit from the enterprise. It has, however, been conclusively shown that the market cannot absorb more than a certain quantity of these products; and that any over-production brings loss and disappointment to the investor. Beside the three kinds of mint above mentioned, there is another species, Mentha arvensis, a native of Japan, which is extensively cultivated there, and is the chief source of the menthol of commerce, so well known as an efficient remedy for neuralgia, migraine, etc., in form of menthol cones. The three varieties of the mint oils previously mentioned are distinguished, aside from their pleasant odor, by the property of leaving a very refreshing and cooling taste in the mouth, and for this reason they form the most important constituent of all fine mouth washes.

True oil of peppermint, Oleum Menthæ piperitæ, when pure is colorless, very mobile, of a burning sharp taste which is followed by a peculiar coolness. The commercial product is usually pale green. Oil of crispmint, Oleum Menthæ crispæ, which in Europe is often sold to novices as oil of peppermint, has always a more or less yellow color and resembles the oil of peppermint in its properties, but it is less fine and cheaper. The same is true of the oil of spearmint, but this has a very characteristic odor and taste, distinctly different from peppermint.

As above stated, the oils of mint are extensively used for mouth washes, also for scenting soap, in liqueurs and pastils, but rarely in handkerchief perfumes.

Oils of Mace and Nutmeg (Oleum Macidis and Oleum Myristicæ).

These oils are prepared either from the seed coat (Oleum Macidis) or the nutmeg itself (Oleum Myristicæ). Oil of mace generally has a yellowish-red color in tint varying from dark to light and even colorless. Its taste is agreeable and mild and the odor exceedingly strong. Like oil of nutmeg, it is extensively used in the manfuacture of liqueurs and for scenting soap. The oil prepared by distillation from the nutmeg is, when fresh, almost colorless or at most faintly yellow, of a burning sharp taste, and an aromatic odor. Like oil of mace, it is used in the manufacture of liqueurs and soaps and also in many perfumes.

In India a third valuable product is obtained from the nutmeg by expression of the ripe fruits and is called nutmeg butter. This is bright yellow and consists of a true fat and an essential oil. Its odor is very pleasant and a very

9

superior soap can be made by saponification of this valuable product with soda lye.

OIL OF MYRTLE.

This oil is of a greenish color and very mobile, but it is not a commercial product; the manufacturer must prepare the oil himself from the leaves, though the yield is small. The articles sold as so-called essence of myrtle are always mixtures of different odors. Southern France, however, exports at high prices a myrtle water (eau des anges) which is really made by distillation of the leaves with water.

OIL OF NARCISSUS.

As to the odor to which this flower owes its fragrance we may repeat what we have said just now with reference to the oil of myrtle: we have never succeeded in obtaining this oil in commerce. The so-called essence of narcissus, though a very pleasant mixture, contains no trace of the true oil. As to

OIL OF PINK,

the same remark applies: the compositions sold under the name of essence d'œillet, however, have a very striking odor of pink.

OIL OF CLOVE (OLEUM CARYOPHYLLI).

This oil when fresh is colorless, but soon becomes yellowish or brown. It is heavier than water in which it sinks and is characterized by an exceedingly strong burning taste and a spicy odor. It remains at least partly fluid at a very low temperature, namely, −20° C. (−4° F.).

OIL OF ORANGE FLOWERS (OLEUM NAPHÆ, OLEUM NEROLI),

commercially known also under the French names huile de fleurs d'oranges, huile néroli, huile néroli pétale, is obtained from the flowers of the orange-tree in Southern France, where the orange is specially planted for this purpose.

This oil is one of the most valuable and most expensive used by the perfumer and is an essential constituent of many of the finest perfumes. Its high cost renders it peculiarly liable to adulteration and it may be stated with certainty that unless the user obtains it from a firm of such a standing as to guarantee the quality and pays a high price for it he has little chance of obtaining a pure product. To add to the difficulties of the situation, there are no chemical tests which may be applied with certainty and dependence must be placed on the nose test. The odor of the purest oil is not as superficially pleasant as that of the adulterated grades as a rule, but in compositions it permits effects which are entirely impossible with the cheaper, adulterated grades.

The French manufacturers of this oil, which is of great importance in perfumery, distinguish several varieties. The most valuable is the oil from the flowers of Citrus vulgaris (or Citrus Bigaradia), the true bitter orange (or Seville orange) tree. This is the so-called neroli bigarade. That called neroli pétale is obtained from the same flowers carefully deprived of their floral envelopes, so that only the petals are subjected to distillation. Much cheaper than these two is the oil of petitgrain which is distilled from the leaves and sometimes also unripe fruits of various trees of the Citrus order.

Oil of Orange.

There are two distinct varieties of orange oil, Oil Orange, Bitter, and Oil Orange, Sweet, the latter being also known as Oil of Portugal. Both varieties are extracted from the peels of the corresponding fruits by mechanical means, the best quality being obtained by hand expression while a poorer grade is made by machine expression, frequently followed by distillation. For the perfumers' use the Oil Orange, Bitter or Bigarade, is the only one of importance, the other variety being used in flavors and beverages. The Italian oil is highest in price and most favored, but a quite satisfactory oil is now obtained from the West Indies.

OIL OF PATCHOULY.

This oil, which might be manufactured with advantage in India, the home of the plant, is, strange to say, not imported from that country, but is distilled in Europe from the dried herb. Fresh oil of patchouly is brown in color, very viscid, almost like balsam, and surpasses all other essential oils in the intensity of its odor. Owing to the strong odor, pure oil of patchouly must really be called ill-smelling; only when highly diluted does the odor become pleasant, and then forms a useful ingredient of many perfumes as the fundamental odor in the harmony.

OIL OF SYRINGA.

Oil of false jasmine, from the flowers of Philadelphus coronarius, is not made as such; in Southern France, however, the flowers are frequently used for the preparation of a cheap pomade known commercially as orange-flower pomade. A personal experiment made with the view to obtain the pure odor by extraction of the flowers with petroleum ether has shown that this plant is suitable for making very fine preparations, both handkerchief perfumes and pomades.

OIL OF PETITGRAIN.

This is the oil obtained by the distillation of the leaves and twigs of the orange-tree, the unripe fruits sometimes being included but resulting in a lower quality of oil. Oil Petitgrain comes mainly from France and Italy but a cheaper quality is obtained from Paraguay. The oil resembles Oil Neroli in odor though it of course is lacking in the delicacy of the latter odor and is of far less value, being used as a stretcher for Oil Neroli in some instances and as a substitute for it in cheap perfumes where the cost of the real oil forbids its employment.

OIL OF RUE (OLEUM RUTÆ).

This oil, obtained by distillation of the herb, is colorless or pale yellow, of a very strong, penetrating odor; it is used in some washes, but more particularly as an ingredient in the manufacture of artificial cognac, for which purpose the plant is specially cultivated in France.

OIL OF RESEDA (MIGNONETTE).

The delightful odor of this plant which formerly could only be fixed by maceration in fat may be readily prepared by extraction with petroleum ether. Yet special precautions should be taken that nothing but portions of the flowers, carefully picked off, and no green leaves are extracted. The oil thus obtained has a yellow color and a disagreeable odor which changes into the well-known pleasant smell of the flower when highly diluted with alcohol.

OIL OF ROSE (OLEUM ROSÆ).

Oil of Rose, more commonly referred to as Otto of Rose, is made in several localities, though as mentioned earlier the finest comes from Bulgaria. Adulteration is prevalent, frequently taking place at the locality of origin and sometimes by the simple expedient of moistening the petals with the adulterant before distillation. This sort of adulteration, which is cleverly calculated not to affect the physical constants of the otto, is difficult of detection and in the main reliance must be placed upon the reputation of the producer and the seller.

True Otto of Rose is frequently slightly greenish in color and of a consistency varying from that of a viscous liquid to one more resembling butter. It becomes almost solid at 14° to 20° C. and even at higher temperatures there is nearly always a separation of a white, paraffin-like substance. The solid constituents of Otto of Rose, which are nearly odorless, are known as stearoptenes and the percentage present is one criterion of purity though not a decisive one on account of the ease with which paraffin or similar substances may be added to a stretched oil.

Otto of Rose is obtained by distillation but the extraction method has been applied to roses in France, resulting in the preparation of the valuable liquid flower essence which is gaining a wide popularity on account of its fidelity to the true rose odor. Rose pomades are also prepared though not as widely used as in the past when the floressence was not available.

Oil of Rosemary (Oleum Rosmarini).

This oil is obtained by distillation from the herb of the rosemary plant as a thin, pale green fluid with an aromatic odor and spicy taste. It is used as an ingredient in some old renowned handkerchief perfumes—for instance, Cologne water—also for flavoring soaps and liqueurs.

Oil of Sage (Oleum Salviæ),

from the flowers of Salvia officinalis, is yellowish, with an odor somewhat similar to that of oil of peppermint, but far less intense. Like the latter, it imparts a pleasant coolness to the mouth and hence is used in some mouth washes.

Oil of Rhodium.

This oil is better known as Oil of Rosewood or Bois de Rose Fermelle. It is a light yellow oil, obtained by the distillation of the wood of Convolvulus Scoparius. Formerly of little importance it is now manufactured in large amounts on account of its high content of linalol, now used in the manufacture of synthetic linalyl acetate, which is an important perfume material aside from its use in stretching Oil Lavender. Oil Rosewood resembles Oil Linaloe, which likewise contains a large amount of linalol, as the name indicates. Neither oil is largely used as such in perfumes except occasionally in the cheaper sorts.

Formerly it was sometimes used for the adulteration of oil of rose, but can also very well be used alone for several perfumes and fumigating preparations.

OIL OF SASSAFRAS (OLEUM SASSAFRAS)

is yellow, spicy, with a burning odor and taste; in the cold it crystallizes only in part. The odor of this oil recalls that of fennel. The purest form of it, or rather substitute for it, is safrol, its main constituent, which is, however, now extracted more economically from crude oil of camphor, in which it likewise forms an ingredient.

OIL OF MEADOWSWEET (OLEUM SPIRÆÆ).

Several species of Spiræa, and especially Spiræa ulmaria, furnish very pleasant odors. This oil consists mainly of salicylic aldehyde.

Despite its pleasant odor and the facility of its production, this substance has thus far found little application in perfumery. The natural oil of meadowsweet, owing to its extremely high price, can hardly ever be used.

OIL OF STAR-ANISE (OLEUM ANISI STELLATI; OLEUM ILLICII)

resembles in its properties the oil of anise, even in its odor; but all connoisseurs agree that the odor of the oil of star-anise far surpasses that of the oil of anise, hence the former is used especially for fine perfumes. This preference, however, does not extend to all preparations. For certain liqueurs, such as anisette, the oil obtained from common anise (Saxon anise) is usually preferred. Many also regard the odor of star-anise as inferior to that of fine European anise.

OILS OF THYME (OLEUM THYMI).

The essential oils of thyme (chiefly Thymus vulgaris) and some related plants are very frequently used for scenting cheap soaps. The oils of these plants are light yellow, and so similar in odor that it is not possible to distinguish them except by direct comparisons.

OIL OF VANILLA.

In the strict sense there is no essential oil of vanilla, the flavoring principle being a crystalline substance, vanillin, which is present to the extent of about 2% in the best grades of Mexican beans. It has a melting point of 76° C. Vanillin is now made artificially from the eugenol of oil of cloves as well as by other methods and is equivalent to approximately forty times its weight of vanilla beans. It is pure white and crystalline, and represents the flavoring principle of vanilla beans, entirely free from inert material.

OIL OF VERBENA

is yellow, with a very pleasant odor of lemons. Its price being quite high, it is usually adulterated with oil of lemon-grass, or else the latter is sold under the name of oil of verbena. In fact the odors of the two oils are so similar that they are easily confounded.

OIL OF VETIVER (OLEUM IVARANCHUSÆ),

from Andropogon muricatus, is viscid, reddish-brown, with a very strong and lasting odor. That obtained from the East Indies is of the best quality and is important to the perfumer.

OIL OF WINTERGREEN (OLEUM GAULTHERIÆ).

This product is obtained by distillation from the leaves and twigs of Gaultheria procumbens or else by distilling the bark or leaves of Betula lenta with water, in which case the oil is generated by the action of the water, as it does not pre-exist in the birch, and, moreover, in this case the oil consists of nothing but methyl salicylate. It differs, like oil of meadowsweet, very markedly from the other aromatic substances and mainly consists of a so-called compound ether. It is a salicylate of methyl, boils at 220° C. (428° F.), is much heavier than water (specific gravity 1.173 to 1.184), and dissolves readily in alcohol and other solvents. It is used chiefly for scenting soap; the perfumes sold as wintergreen are usually mixtures of different substances which contain no oil of wintergreen.

OIL OF YLANG-YLANG (OLEUM UNONÆ ODORATISSIMÆ).

This oil is imported from Manila, Madagascar and the Bourbon Islands, the value of the oil being in the order named. It is a nearly colorless oil of pleasant odor and of great value as a constituent of some of the most popular odors. French perfumers use it largely but Americans have failed to appreciate its value fully. It is subject to adulteration and the brand is the best protection to the buyer who is unwise to trust himself to an oil of doubtful quality. Adulteration is difficult to detect except by the nose test which, of course, requires experience and skill.

OILS OF CINNAMON (OLEUM CINNAMONI).

Commercially we find chiefly three varieties of essential oils which are designated as: oil of Ceylon cinnamon, oil of Chinese cinnamon or oil of cassia, and oil of cinnamon leaves. Oil of Ceylon cinnamon, sometimes called "true oil of cinnamon," made from the bark of the twigs of the cinnamon laurel and formerly imported mainly from Ceylon but now distilled in large amounts in Germany from imported cinnamon "chips," is rather viscid, golden yellow to reddish-brown in color, of a burning though sweet taste. In the air it gradually absorbs oxgyen, when it becomes dark red, thicker, and of weaker flavor. Oil of Ceylon cinnamon, which should always be used in perfumes or liqueurs when simply "oil of cinnamon" is directed, has a specific gravity of 1.030 to 1.035 at 15° C. (59° F.) and boils at about 240° C. (464° F.). Its chief constituent upon which its aroma depends is cinnamyl aldehyde.

Oil of Chinese cinnamon, or oil of cassia, has for a very long time, up to within a few years, always reached the market in a more or less adulterated state, a regular practice of the Chinese exporters being to dissolve ordinary resin in it (claiming afterward that the "resin" was caused by the oxidation of the oil through age) and often also to add petroleum to it. These frauds have been well shown up by Schimmel & Co., of Leipsic; and in consequence thereof, the quality of oil of cassia exported from China has been greatly improved. Oil of cassia when pure has a specific gravity of 1.060 to 1.065, and should contain not less than seventy-five per cent of cinnamyl aldehyde.

Oil of cinnamon leaves is an inferior product, often used for adulterating oil of Ceylon cinnamon. It does not deserve notice by the perfumer.

As an appendix we may add in this connection a description of the

Oil of Turpentine (Oleum Terebinthinæ),

because it must be called an important substance to know for the perfumer, inasmuch as it is very frequently used for the adulteration of different essential oils. Oil of turpentine, which is obtained from incisions into the bark of different fir and pine trees, the exuding resin being distilled with water, comes into commerce from various sources. Different sorts are distinguished, but to the perfumer only the rectified oil of turpentine, oleum terebinthinæ rectificatum, is important. Oil of turpentine has a yellowish color and a decidedly disagreeable, resinous, and burnt taste. By repeated distillation, especially over quicklime or chloride of lime (bleaching powder), it is finally obtained as a colorless, very refractive liquid with a density of 0.855 to 0.870 and a boiling-point at 160° C. (320° F.). Its odor is peculiar, but not easily distinguished from that of old essential oils, such as oils of caraway, anise, etc. One peculiarity of oil of turpentine is that its odor is easily masked by that of other essential oils, so that, for instance, a comparatively large quantity of oil of turpentine needs the addition of but little oil of anise to impart to the entire mixture a rather pronounced odor of anise. This peculiarity has led to the frequent employment of rectified oil of turpentine for the adulteration of other essential oils.

CHAPTER IX.

THE ADULTERATIONS OF ESSENTIAL OILS AND THEIR RECOGNITION.

WE find it necessary to devote a special chapter to the adulterations of the commercial essential oils because an experience of many years has shown us that hardly any other group of products is subject to so many sophistications as essential oils. The high price of most aromatic substances and the difficulty of recognizing the adulteration furnish an inviting field to the unscrupulous manufacturer. In the best interest of the perfumer, therefore, we advise the purchase of essential oils only from renowned reliable houses, even at higher prices, for the cheap commercial products are almost worthless, since they are almost without exception adulterated.

The adulterations are very manifold. Some expensive oils are mixed with cheaper ones having a similar odor—for instance, oil of rose with oil of geranium or oil of geranium grass; oil of orange flowers with the oil from Philadelphus coronarius; oil of verbena with oil of lemon grass; oils of caraway, anise, and fennel with oil of turpentine; oil of cinnamon with oil of cassia, etc. Besides these, other deceptions are practised—for instance, oil of anise is mixed with oil of turpentine and in order to make the mixture congeal readily (which is the characteristic of true oil of anise, as above stated) paraffin or spermaceti is added. A similar practice prevails with adulterated oil of rose and other viscid oils. Oil of bitter almond we have found adulterated with or entirely replaced by nitrobenzol, etc.

The demonstration of the adulteration of an essential oil by chemical means offers many difficulties. We devote particular attention to the physical characteristics, for experience has shown us that the olfactory organ—provided it is very expert—is often able to determine the genuineness of any aromatic substance when other tests have given only uncertain results, or can give certain results only in the hands of experts. To make this test, however, quite reliable, it is necessary to be familiar with the substances in their pure unadulterated condition.

The manufacturer of perfumery, therefore, should spare neither trouble nor pecuniary sacrifices to obtain possession of absolutely genuine specimens of those essential oils, even in minute quantities, which he intends to employ. Such samples should be carefully preserved (protected from heat, evaporation, daylight, etc.) for the purpose of immediate comparison with the oils to be purchased.

As above stated, the physical properties of the essential oils usually furnish the means of recognizing their purity, and these give more reliable results to the practical perfumer than the chemical tests. The most valuable points are furnished by the boiling-point, the congealing-point, and the density of the oils. The following table gives the boiling and congealing points of the most important essential oils in degrees of the centigrade thermometer, together with the density (or specific gravity); where two figures are given, they indicate the extreme limits found in genuine samples.

Special characteristics of some essential oils with reference to their action at low temperatures or their melting-point are given in the column " Remarks."

Oil of turpentine, paraffin, wax, and spermaceti being frequently used for the adulteration of essential oils, have been included in the table.

If accurate results are aimed at in the examination of an essential oil according to this table, the specific gravity should be determined by means of a scale sensitive to one one-thousandth gram, and the thermometer should be graduated to the tenth of a degree.

TABLE SHOWING THE APPROXIMATE DENSITY, BOILING AND CONGEALING POINTS OF THE MOST IMPORTANT ESSENTIAL OILS USED IN PERFUMERY.

Essential Oil of	Density.	Boiling-Point, Deg. C.	Congealing-Point, Deg. C.	Remarks.
Absinth................	0.895	.....	.....	
Anise................ ...	0.980	.....	+10-15	
Bergamot	0.850–0.890	188	−24	
Bitter almond	1.040	180	.....	
Do., art. (nitrobenzol)..	1.866	213	+3	
Cajuput................	0.880	.....	.....	
Calamus	0.962	.....	.....	
Camomile	0.924	160–210	.....	
Camphor (Borneo).	.....	212	.....	Melts at 198
" (Chinese)......	0.985	205	.....	Melts at 175
Caraway...............	0.960	195	.....	
Cassia...	1.060	252–255	.....	
Cedar wood............	....	264	−22	
Cinnamon..............	1.030–1.035	240	below−25	
" leaf..........	1.053	.....	.....	
Clove................ ...	1.034–1.055	248	below 20	Forms crystals −16
Coriander..............	0.871	150–200	.....	

Essential Oil of	Density.	Boiling-Point, Deg. C.	Congealing-Point, Deg. C.	Remarks.
Crispmint...........	0.978	.....	.....	
Cubeb...............	0.880	.....	.....	
Fennel	0.960–0.980	.....	+8	
Gaultheria	1.173	224	.....	
Geranium....	0.895	216–220	.....	Forms crystals — 16
Hyssop............	0.889	.....	.....	
Juniper............	0.870	.....	.. ..	
Lavender	0.870–0.940	186–192	.....	
Spike-lavender.....	.....	140	.....	
Lemon	0.850–0.870	177–250	.....	
" grass........	0.870–0.898	220	−22	
Limetta.	0.931	.....	.....	
Mace	0.890–0.950	.....	.....	
Marjoram.....	0.890–0.920	163	.....	
Melissa............	0.855	.....	.....	
Neroli.	0.889–0.889	175	.....	Forms crystals — 16
Nutmeg...........	0.880–0.948	172	.....	
" butter......	.....	.....	31	
Olibanum..........	.....	162	.....	
Orange, bitter..... ...	0.830–0.860	176	.....	
" sweet.......	0.840–0.850	176	. ...	
Parsley............	1.015	.....	.....	
Patchouly..........	0.950–1.012	282–294	.....	
Peppermint	0.902–0.930	188–212	.....	
Portugal (orange peel)...	0.840–0.850	176	.....	
Rose...........	0.832	229	+14–20	
Rosemary	0.895–0.916	185	.....	
Rue................	0.911	.....	.....	
Sage	0.902	.....	.....	
Santal.............	0.950–0.980	288	−22	
Sassafras	1.082	.....	.....	
Serpyllum	0.890–0.920	.....	.....	
Star-anise.....	0.982	.....	.....	
Thyme.............	0.870–0.940	170–180	.....	
Vanilla	.....	150	76	
Vetiver	1.007	286	.....	
Wintergreen.........	1.180	220	.....	
Ylang-ylang.........	0.980	.....	.....	
Turpentine	0.855–0.870	160	.....	
Paraffin	0.870	.....	.....	Melts at 50–65
Wax	0.960–0.970	.....	.....	Melts at 65–70
Spermaceti.........	0.943	.....	.....	Melts at 45–50

In buying essential oils, except it be from a house whose reputation is a guaranty of their genuineness, it is to the interest of the perfumer to make a test. He must look for certain substances which are generally used for the sophistication of essential oils. These are: A. Other essential oils; B. Fixed oils; C. Alcohol; D. Paraffin, spermaceti, wax; E. Other substances, including synthetics.

A. Adulteration of Essential Oils with Other Essential Oils.

This mode of adulteration, which is frequent, is naturally the one most difficult of demonstration. In the case of cheap oils such as those of caraway, lemon, orange peel, etc., rectified oil of turpentine is almost without exception the adulterant. The methods usually recommended, such as attempting to dissolve out the oil of turpentine by strong alcohol, hoping thus to separate it from the essential oil, are without practical value.

The adulteration can, however, often be demonstrated by rubbing a drop of the suspected oil on a glass plate and testing the odor, provided the olfactory organ is trained. As the above table shows, the oils have different high boiling-points, while oil of turpentine boils at a rather low temperature, hence it evaporates sooner than the others and can be demonstrated by its odor.

The demonstration of an adulteration with an essential oil is most certain by so-called fractional distillation. Some of the oil to be examined (about four to six fluidrachms) is placed in a small retort with condenser and heated to a temperature a few degrees below the boiling-point of the oil in question. If, for instance, oil of bergamot adulterated with oil of turpentine is to be tested, it is heated carefully to nearly 188° C. (370° F.), the boiling-point of the oil of bergamot; the oil of turpentine which boils at 160° C. (320° F.) passes over completely, while the oil of bergamot remains in the retort.

B. Adulteration of Essential Oils With Fixed Oils.

This form of adulteration is so crude as to be only occasionally used, since it can be readily detected by fractional distillation, by solubility tests or, most simply, by spotting on paper. The translucent stain characteristic of oils will disappear entirely in a reasonable time if made only by volatile oils while fixed oils leave a stain which does not disappear, even on heating.

C. Adulteration with Alcohol.

This frequent adulteration is demonstrated either by fractional distillation, when the alcohol passes over first between 70° and 80° C. (158° and 176° F.), or by the use of the vessel illustrated in Fig. 31, which is divided into 100 equal parts.

The vessel is filled to the tenth division with the oil to be tested, and water is added to bring the volume to the 50 mark. If alcohol is present, it is taken up by the water so that the volume of oil appears to diminish. If the oil reaches to the mark 7, it contained three volumes of alcohol, or in other words it was mixed with thirty per cent of alcohol.

Fig. 81

It is true, essential oils likewise dissolve somewhat in water, but in such minute quantities as not to affect the success of the test.

D. Adulteration with Paraffin, Spermaceti or Wax.

Only viscid oils are commonly adulterated in this way, particularly oils of rose and anise. Fractional distillation will detect it in most cases but in the case of oil of rose it is advisable to subject the residue to chemical tests to prove its nature as the stearoptenes are likewise difficultly volatile. They are, however, readily oxidizable, while paraffins are not.

E. Adulteration with Other Substances.

Earlier methods of adulteration were crude and the advancement of chemical science made them easy to detect and they were for the most part abandoned. But chemical science, while making the detection of the old methods easy, provided a basis for scientific adulteration which aims at duplicating the physical and chemical characteristics of the pure oil as accurately as possible. Needless to say, such adulterations are not easy to detect.

In cases where oils are judged by the acetyl numbers or the ester numbers, i. e., by tests showing the percentage content of alcohols or esters, the oil is stretched with an inert material or with poorer oils and then the requisite amount of some alcohol or ester is added to bring the analysis up to the proper figures. Let us take a specific example. French Oil Lavender is stretched with Oil Aspic which contains practically no esters. Then a certain amount of odorless ethyl citrate or ethyl phthalate is added to produce the correct ester number on analysis. Sometimes a better product is made by the addition of synthetic linalyl acetate which no chemical analysis can differentiate from the linalyl acetate of pure Oil Lavender.

In a similar way the washed terpenes of Oil Lemon, worthless from the perfume standpoint, may be treated with the proper amount of citral, producing a fake Oil Lemon. Oil of Bitter Almonds may be and often is adulterated with benzaldehyde, its principal constituent, which can be made far more cheaply by chemical means. Even benzaldehyde may be sometimes adulterated with the much cheaper Oil of Mirbane or nitrobenzol, a fraud which can be easily exposed by the fact that benzaldehyde is perfectly soluble in a solution of sodium bisulphite.

The detection of these clever chemical adulterations requires a careful study of each case, with the application of the most suitable tests, and is frequently almost impossible except by the most delicate test, that of odor value. Usually, however, even the cleverest adulteration will affect some one of the physical constants and even when it is not practical to identify the adulterant sufficient evidence can be obtained to indicate that fraud has been committed.

CHAPTER X.

SYNTHETIC PRODUCTS.

THE subject of artificial or synthetic perfumes is of surpassing interest and importance to the practical perfumer as he is constantly confronted with the consideration of synthetics, new and old, and their usefulness and practical value as compared with natural products. In particular, the saving of expense made possible by their use is sufficient to warrant their employment wherever possible. In order to comprehend their real field of usefulness, it is first necessary to ascertain where the line between natural and synthetic products should be drawn.

By natural products we mean those which are directly drawn from flowers, leaves, roots, herbs, fruits, etc., whether by expression, distillation, maceration, or extraction by volatile solvents.

By what are known as synthetic products are meant those products which are obtained from a natural base by chemical treatment or by combination of one or more organic or inorganic chemicals to produce the aromatic or odorous bodies. The resulting combinations are more or less contaminated with impurities. These are removed by purification and it is on the care and skill with which this purification is effected that the quality and odor value of the synthetic depend.

It should be borne in mind that many of the synthetics are made indirectly from vegetable bases, for example: Heliotropin from oil of camphor, vanillin from oil of cloves, terpineol from oil of turpentine, the artificial violets from oil of lemongrass. These few instances of products so universally used should serve to dispel one of the general misconceptions that most, if not all, of the synthetics are coal-tar derivatives.

Nothing was ever further from the truth, as any perfume chemist knows, yet it is often difficult to correct this false conception. It is therefore plain that there is some overlapping of the natural products and the synthetic, in those instances in which the synthetic is derived from vegetable bases, but where this derivation is only a metamorphosis of the needed chemical elements, and not of the odorous principle as a whole, they may be termed synthetic.

It may be primary to say that synthesis in this field, as in all chemistry, is based upon analysis, but it is most necessary to emphasize this if the proper understanding is to be had. It was only after the pure natural products had been carefully analyzed and their elements found that it was possible to search for these elements in other less costly substances of vegetable or mineral origin, and recombine them into artificial aromatics, resembling the compounds sought to be reproduced. And yet, up to the present time, this work of analysis and synthesis leaves much to be desired. Not yet has it proved practical to imitate the delicate chemistry of nature in all of its infinite detail, and for this very reason it is not yet possible, if it ever will be, to reproduce scientifically and successfully all of the characteristic qualities of any natural product.

In fact, it seems as if the natural and synthetic perfume materials supplement one another most happily. It is a provable fact that in the process of extracting natural products some important odorous elements are lost, and it is these missing notes necessary to the complete chords of the odorous harmony which are sometimes supplied by well-made synthetics. It is indisputable that many synthetic products often intensify the odors of the natural products.

The development of artificial perfumers' materials has made possible the manufacture of many new and novel odors, as well as the perfecting of the old staple odors. Concurrently with the marked advances made by the manufacturers of synthetics during the last twenty years or more, the manufacturers of the natural products have not been unprogressive. They have expended fortunes upon experiments in fertilizing the soil, crossing of types, and the grafting of plants, at the same time that they have perfected their methods of distillation and extraction, heightened the efficiency of their machinery, and secured the ripest results of expert chemical aid, so as to secure all that was possible from the crude materials; and by the discovery of new solvents they have succeeded in obtaining products of great strength and purity. In this way the wide use of synthetics has had an excellent effect in stimulating the manufacturers of the natural products to improve their processes in every direction.

Regardless of the fact that the consumption of synthetics has increased enormously, as they deserved to increase with the spread of knowledge and appreciation of the value of these important ingredients (they are to-day indispensable in every perfumer's laboratory), it would be taking an unwarranted liberty with the truth to state that the synthetics can replace the natural products. These are still the real basis of all good perfumery, and are just as necessary as ever for the attainment of many perfume effects which are impossible otherwise. In the proper combination of the natural and synthetic products, based upon a clear understanding of the proper functions of each in the economy of the perfume industry, lies the best guarantee of success. Each has its work to do; neither is complete without the other, and working harmoniously the results will be both rich and satisfying.

The development of synthetic perfume materials is still in its infancy and it may confidently be expected that the future will see many new and useful products added to the list now available. The list which follows this chapter gives most of the important ones now in use together with their physical constants but cannot claim to be absolutely complete, owing to the fact that new products are appearing from time to time, frequently under names which disguise their chemical nature and do not serve to indicate whether the products in question are true chemical identities or mixtures.

With the coming of the synthetics the perfumer has had a new set of difficult problems to solve. The flower perfumes are themselves nicely blended by nature to produce desirable odors, the varying notes being delicately harmonized. The perfumer, therefore, in using these has merely to mix them in easily ascertainable proportions and can hardly fail of producing a reasonably pleasing odor. This is far from true of the synthetics, each of which, assuming the impossible case that they are absolutely pure, represents but a single note in the scale of odor harmony. The blending of these, in themselves often disagreeable odors, to produce a delightful composition is a task of real magnitude and much progress is still possible. At best it is practically impossible to duplicate with exactitude the fine delicacy of the flower perfumes and it is for this reason that the synthetics can never entirely replace the natural perfumes, whatever their advantage in the direction of economy.

The correct use of these materials will be taken up as fully as is practical in an added chapter dealing not only with the synthetics themselves but with the many valuable specialties which have been compounded by the manufacturers for the convenience of perfumers.

CHAPTER XI.

Table of Physical Characters of Aromatic Chemicals.

By John C. Umney, F.C.S., London, England.

Products	Melting Point	Boiling Point	Specific Gravity	Rotation	Refractive Index
HYDROCARBONS—					
Styrol................		140°	0.9074		1.54344
Cymol................		173.5°	0.8595		1.48456
Pinene...............		155–156°	0.858		1.4655
Camphene............	51–52°	158–159°	0.842(54°)		1.45564 (54°)
Limonene............		175–176°	0.846		1.47459
ALCOHOLS—					
Citronellol..........		117–118° (17mm)	0.856	+4°0′	1.45789(l) 1.45659(d)
Androl..............		197–198°	0.858	−7°10′	1.44991
Geraniol.............		226–227°	0.882	±0°	1.4773(15°)
Nerol...............		225–227°	0.866	±0°	
Linalol.............		194–198°	0.867(20°)	from −8° to −9°10′	1.4671(20°)
Borneol.............	203–204°	212°	1.02	from +37° 44′ to +37° 77′	
Isoborneol..........	212°				
Terpineol...........	35°	218°	0.9057	from +95° 9′ to −117°5′	1.48084
Pulegol.............		215°	0.912(10°)		1.4792
Menthol............	42°	211–212°	0.890(20°)	−43°45′	1.4479
Cinnamic alcohol.....	33°	257°	1.03		
Phenylethylic alcohol...		220–222°	1.023(15°)		1.532
Myrtenol............		220–221°	0.985	+49°25′	
Octylic alcohol........		196–197°	0.8278		
Nonylic alcohol.......		213.5°	0.840(15°)	±0°	1.43582
Undecylic alcohol.....		231–233°		−1°18′	
Benzylic alcohol......		205°	1.05		1.540
Phenylpropylic alcohol..		234–235°	1.008		
ALDEHYDES—					
Citral...............		228–229°	0.8977(15°)		1.4931
Citronellal..........		205–208°	0.8538(17°)	from +10° to +11°	1.4481
Benzoic aldehyde......		177°	1.050(15°)	±0°	1.545
Cinnamic aldehyde.....		128–130° (20mm)	1.054	±0°	1.6194
Salicylic aldehyde......		197°	1.1698		
Anisic aldehyde.......		248°	1.122	±0°	1.572
Octylic aldehyde.......		60–61° (9mm)	0.8211		1.41955
Nonylic aldehyde......		80–82° (13mm)	0.8277		1.42752
Decylic aldehyde.......		207–209°	0.828(15°)		1.42977
Vanillin..............	80–81°	170° (15mm)			
Heliotropin...........	37°	263°			
Phenylacetaldehyde.....		205–207°	1.085	±0°	1.52536
Furfurol.............		160.5°	1.1594		

TABLE OF PHYSICAL CHARACTERS OF AROMATIC CHEMICALS.
(Continued.)

Products	Melting Point	Boiling Point	Specific Gravity	Rotation	Refractive Index
ETONES—					
Methylamyl ketone.....		151–152°	0.8366		
Ethylamyl ketone......		169–170°	0.8254		1.41535
Methylheptyl ketone....		194–196°	0.83178		
Methylhexyl ketone....		172–173°	0.8185		
Methylnonyl ketone....	13.5°	224–226°	0.8268		
Methylheptenone.......		173–174°	0.8602		1.44003
Carvone...............		223°	0.9598	+59°57′	1.49952
Fenone....·........	5–6°	192–193°	0.9465		1.46306
Camphor..............	175–176°	209°	0.9853	±44°22′	
Pulegone.............		221–223°	0.9223		1.47018
Isopulegone..........		90° (12mm)	0.9213	+10°15′	1.4690
Menthone.............		206–207°	0.8960(20°)	from +21°16′ to +24.78′	1.4496
Ionone a.............		123–124° (11mm)	0.932(20°)		1.50001
" b.............		127–128° (10mm)	0.946		1.521
Acetophenone.........	20.5°	202°	1.032(15°)		
Jasmone..............		257–258°	0.945(15°)		
Irone................		144° (16)	0.937(20°)	+40°	1.50113
Methylionone a........		140–150° (20)	0.925–0.931		
" b.......		140–155° (20)	0.935–0.938		
STERS—					
Acetate of linalyl......		105–108° (11mm)	0.912(15°)		1.451–1.454
" " geranyl......		127–129° (16mm)	0.9174(15°)	±0′	1.462–1.464
" " citronellyl..·.		119–121° (15mm)	0.8928(17°)	+2°37′	1.4456
" " bornyl.......		106–107° (15mm)	0.991(15°)	−38°2′	1.46635
" " terpinyl.....		90–94° (5mm)	0.9544(15°)	±0′	1.46505
" " benzyl......		206°	1.057(16°)		
Salicylate of methyl....		224°	1.1969(0°)		
Benzoate of benzyl.....	21°	323–324°	1.1224		
Anthranilate of methyl..	24–25°	132° (14mm)	1.168		
Butyrate of methyl.....		103°	0.91939(0°)		
" " ethyl.......		119–120°	0.89957(0°)		
Isovaleriante of ethyl...		134°	0.88514(0°)		
Caprylate of ethyl.....		207–208°	0.8871		
Valerianate of citronellyl		194–196°			
Formate of terpenyl.....		135–138° (40mm)	0.9986(0°)	−69°25′	
Cinnamate of methyl...	33.4°	263°	1.0415		
" " ethyl.....	12°	271°	1.0498(20°)		
Salicylate of amyl.....		151–152° (15mm)	1.049–1.055	+2°	1.505
" " benzyl....					

TABLE OF PHYSICAL CHARACTERS OF AROMATIC CHEMICALS.
(Continued.)

Products	Melting Point	Boiling Point	Specific Gravity	Rotation	Refractive Index
ESTERS (Continued) –					
Acetate of amyl........		138–139°	0.8837		
"　" phenylethyl..		232°	1.038(15°)		
"　" neryl........		134° (25mm)	0.917(15°)		
"　" menthyl.....		244°	0.985(20°)		
"　" cinnamyl....		135° (11mm)			
"　" isobornyl....		107° (13mm)	0.9905		
Salicylate of ethyl......		233°	1.1843		
"　" phenylethyl					
Benzoate of amyl.......		260°	1.0034		
Anthranilate of ethyl....	120°	260°			
Butyrate of amyl.......		178–179°	0.8823(0°)		
Anisate of methyl......	45–46°	255°			
Myristate of ethyl......	10–11°	295			
Triacetate of glycerin...		258–259°	1.155		
Iridate of methyl.......		366°			
Formate of geranyl.....		112–114° (25mm)			
"　" neryl.......		119–121° (25mm)			
"　" linalyl......		189–192°			
"　" bornyl.....		100° (14mm)	1.017(15°)		
"　" methyl.....	9°	219°	0.9444		
Acetate of menthyl.....		224°	0.985	−79°42′	
ACETALS—					
Methylephenylglycol....		218°	1.1334(6°)		
Ethyldenphenylglycol...		222°	1.062		
Isobutylidenphenylglycol		126° (15mm)	1.027(15°)	•	
Isovaleridenphenylglycol		148° (20mm)	1.038(18°)		
LACTONES—					
Coumarin.............	67°	290–291°			
Methyl-coumarin.......	90°	292–293°			
Ethyl-coumarin........	70–71°	299°			
PHENOLS AND ETHERS OF PHENOLS—					
Thymol..............	50–51°	232°	0.9760(15°)		1.52269
Carvacrol............		236–237°	0.976(20°)	±0′	1.523
Eugenol.............		252°	1.0713		1.541
Iso-eugenol...........		258–262°	1.080(16°)		1.5680
Methyl-eugenol.......		248–249°	1.041(11°)		1.5373
Aceto-eugenol........	29°	281–282°	1.0842(15°)		
Benzyliso-eugenol......	58–59°				
Methyliso-eugenol......		263°	1.064		1.5720
Safrol...............	11°	233°	1.108		1.53836
Isosafrol.............		253–254°	1.124–1.129		1.580
Apiol...............	30°	294°	1.1788		1.5380
Isoapiol.............	55–56°	304°			
Anethol.............	21°		0.986		1.56149
Estragol.............		97–98° (12mm)	0.972(15°)		1.52355

TABLE OF PHYSICAL CHARACTERS OF AROMATIC CHEMICALS.
(Continued.)

Products	Melting Point	Boiling Point	Specific Gravity	Rotation	Refractive Index
NITRIC COMPOUNDS—					
Nitrobenzene.........		209–210°	1.2002		
Musk (hydrocarbon)....	110°				
" ketone	131°				
" ambrette	84–86°				
NITROGEN-CONTAINING PRODUCTS—					
Indol................	52°	253–254°			
Skatol...............	95°	265–266°			
OXIDES—					
Eucalyptol...........		176°	0.9267(20°)		1.45839
ACIDS—					
Phenylacetic acid.......	76–77°	265°			
Cinnamic acid.........	133°	300°			
Benzoic acid..........	121°	249°			
Citronellic acid.......		257°			
Salicylic acid.........	155–157°				
Anisic acid	104°				

CHAPTER XII.

THE ESSENCES OR EXTRACTS EMPLOYED IN PERFUMERY.

THE term *essence* or *extract* in perfumery means a solution of an aromatic substance in strong alcohol. These solutions are generally made as concentrated as possible and in this form find application in the manufacture of handkerchief perfumes and of certain odors bearing a special name. The so-called extrait d'œillet, extract of pink, or the favorite perfumes known as new-mown hay have nothing in common with either pink or hay except the name; like many other odors, both are merely mixtures of different essences·or extracts.

Besides the manufacture of true perfumes, essences or extracts are also used for scenting fine soaps, sachets, mouth washes, etc. For the latter, too, use is often made of the so-called aromatic waters (eaux aromatisées) which are obtained as a by-product in the distillation of fragrant plants, and have a very fine odor owing to the small amount of the aromatic substance they hold in solution. To this class belong orange-flower water (Aqua Naphæ triplex, eau de fleurs d'oranges), peppermint water (Aqua Menthæ, eau de menthe), and many others.

Essences or extracts can be made in two ways: in the case of aromatic substances which are obtainable in the pure state—that is, essential oils—by dissolving them in strong alcohol in definite proportions; in the case of aromatics combined with a fatty substance by one of the processes described above, by treating the pomade (lard, or other perfectly bland, sweet, and in itself odorless fat combined with the aromatic) or huile antique (fixed oil holding the aromatic substance in solution) with the strongest alcohol.

According to the action of the alcohol upon the pomade or huile antique at ordinary or higher temperature, the process is called cold or warm infusion. Cold infusion furnishes the odor in a much more delicate and superior form than the warm. The cold infusion requires for complete solution of the aromatic four to six weeks; the warm, ten to fourteen days. Although the former consumes a much longer time, it is to be preferred, as the heat injures the odor. Pomades or huiles antiques are never completely exhausted by a single treatment with alcohol. Even when heat is employed they always retain a portion of the aromatic with great tenacity; a second and third infusion still abstracts odor from them, and finally nothing remains but pure fat with a pleasant odor which is stained and sold commercially as pomade under the name of the respective odor—violet, orange flower, reseda, etc.—or else is used over again in the factory for the extraction of flowers.

Experience has shown us that it is best to infuse the pomades or huiles antiques twice in the cold and to use the two fluids united for the finest perfumes; the residue by warm infusion furnishes an essence of second quality, and superior pomades or fragrant oils. The infusion is generally effected in strong glass bottles of a capacity of three to five gallons; about five to six quarts of cologne spirit being poured over six to eight pounds or pints of fat or huile antique.

In treating huiles antiques all parts of the oil should be brought into contact with the alcohol as much as possible, hence the bottles must be frequently shaken; a better plan is to bring the tightly closed bottles into an apparatus in which they are constantly agitated by rotation. Such an apparatus is easily made by placing the bottles in an inclined position between two rods fastened to a common axis which is kept revolving. The adjoining illustration (Fig. 32) shows such a contrivance which is required also in the manufacture of perfumes. The rotation may be effected by clockwork, water power, or any other motor.

Pomades being solid must be divided into small pieces which may be done with a knife, but the following procedure is more suitable and less laborious. The pomade is placed in a tin cylinder four inches wide and about a foot high, which is open at one end, the other being closed with a tin plate having several fine openings. The cylinder filled with pomade is set upon the bottle containing the alcohol for extrac-

FIG. 32.

tion, and the pomade is pressed through the openings in the shape of thin threads by means of a piston.

In this way, of course, the pomade acquires a very large surface and rapidly yields the aromatic substance to the alcohol. The odor of the pomade differs according to the length of time which it has been subjected to the flowers, and on being treated with alcohol furnishes extracts of corresponding strength. This should be borne in mind in the manufacture of perfumes which are intended to be uniform in quality.

After two cold and one warm infusion of the pomade, it may be made to yield some more aromatic material by heating it carefully to its exact melting-point, when extract again

appears on the surface and can be poured off by gentle inclination of the vessel.

In the following pages we give the proportions by weight
and measure employed by the most important French, English, and German manufacturers for their pomade extracts or
solutions of the essential oils in alcohol. As to the latter we
again repeat that it must be over 88 to 90% strength according
to Tralles or even stronger, and that it must be absolutely free
from any trace of amyl alcohol (potato fusel oil), the least
amount of which impairs the delicacy of the odor. In this
country (the United States) there is no difficulty whatever in
obtaining alcohol of proper strength. The market offers
scarcely any other but that of 94%. Of course deodorized
alcohol, or so-called Cologne spirit should be used. Grain
and wine spirits are the kinds which when rectified are to be
preferred to all others. All the citron oils (*i.e.*, oils of lemon,
bergamot, and those with similar odor), rose oils (oils of rose,
geranium, and rhodium), and many other sweet scents are
most fragrant when dissolved in pure spirit of wine, while the
odors from the animal kingdom and those of violet (violet
and orris root) smell sweetest when dissolved in grain spirit.

The essences prepared from pomades or huiles antiques
usually contain in solution some fat which is best removed by
cooling. To this end the vessels containing the essences are
placed in a vat and surrounded with pellets of ice and crystals
of chloride of calcium. By this mixture the temperature can
be reduced below − 20° C. (− 4° F.), and after some time the
fats are deposited in a solid form at the bottom of the vessel.
This is then taken from the vat and the essence carefully
poured from the sediment.

The alcoholic extracts of the pomades or solutions of the
aromatics are called essences or extracts (French, extraits);
the solutions obtained from resins and balsams are usually
termed tinctures.

While some extracts, owing to their strong odor, can be used only when diluted with alcohol, others are employed in perfumes as such. Pure extracts (extraits purs) are those containing only a single odor and are but rarely used as perfumes; the latter are usually mixtures of several, often a great many odors.

CHAPTER XIII.

DIRECTIONS FOR MAKING THE MOST IMPORTANT ESSENCES AND EXTRACTS.

There is always danger of confusion to the reader of books on perfumes in the fact that the different terms have a varying significance in French and English. Unfortunately, inexpert translators have sometimes used the French terms as synonymous with the corresponding English ones. This is best cleared up by defining the most important ones and giving the meanings in both languages.

Taking first the French words, *essence* is used in the same sense in which we use essential oil in English. Thus, *essence de rose* is equivalent to otto (attar) of roses. *Extrait* does not mean the same as our English extract but is applied only to the alcoholic washings of solid concretes. Similar washings of pomades are not correctly referred to as extracts but as *lavages* and it is in this sense that the word is used in French works on the subject. *Esprit* has the same meaning as our English "spirit," being applied to alcoholic solutions of volatile oils, as *esprit de menthe,* etc.

In English the word essence is only infrequently used, being never properly applied to volatile oils, and is confined to the alcoholic solutions of volatile oils. Thus we have essence of peppermint, essence of citronella, etc. It is rarely used by perfumers in any sense. Practically it is equivalent to our term spirit, which likewise is seldom employed by perfumers. Extract is a word which is given a very broad and ill-defined usage in English. It is applied on the one hand to the alcoholic liquids obtained by washing of pomades and solid concretes and on the other is the term most frequently used in connection with finished perfumes to distinguish them from the weaker toilet waters. The exact meaning which is intended must be judged in connection with the context.

Tincture is a term frequently used in English though only occasionally met with in French writings on the subject of perfumes. It is properly applied only to the alcoholic extracts made by the solution of the alcohol-soluble constituents of materials which are incompletely dissolved. Tincture of benzoin and tincture of orris are made by digesting the corresponding substances in alcohol and straining out the insoluble portions. As used by perfumers the word retains practically the same meaning as when applied to drugs.

In making the various tinctures and extracts spoken of in this and succeeding chapters it must be remembered that the best results can be obtained only by the use of the finest Cologne Spirits. The quality of the alcohol used has an important influence on the finished perfume.

TINCTURE OF AMBERGRIS (EXTRAIT D'AMBREGRIS).

Ambergris .... 4 oz.
Alcohol .... 1 gal.

The ambergris should be broken into small pieces with a chopping knife repeatedly moistened with alcohol, and allowed to digest in the alcohol for some weeks at a temperature of about 30° C. (86° F.).

TINCTURE OF BENZOIN (EXTRAIT DE BENJOIN).

Benzoin .... 2 lbs.
Alcohol .... 1 gal.

This tincture is not so much used for handkerchief perfumes as for preserving many pomades, as it possesses the valuable property of preventing fats from becoming rancid.

TINCTURE OF OAKMOSS.

Resinarome oakmoss	4 oz.
Alcohol	4 qts.

TINCTURE OF CASTOR (EXTRAIT DE CASTOREUM).

Castor	4 oz.
Alcohol	1 gal.

TINCTURE OF MUSK SEED (EXTRAIT D'AMBRETTE).

Musk seed	2 lbs.
Alcohol	1 gal.

TINCTURE OF LABDANUM.

Gum labdanum	4 oz.
Alcohol	4 qts.

TINCTURE OF ORRIS ROOT.

Orris root	2 lbs.
Alcohol	4 qts.

TINCTURE OF OLIBANUM.

Gum olibanum (Frankincense).............	2 lbs.
Alcohol	4 qts.

TINCTURE OF STYRAX.

Gum styrax	2 lbs.
Alcohol	4 qts.

TINCTURE OF CEDAR (EXTRAIT DE BOIS DE CÈDRE).

This is made by digesting finely rasped cedar wood with strong alcohol, namely:

Cedar wood chips.......................... 6 lbs.

Alcohol .... 5 qts.

TINCTURE OF CIVET.

Civet .... 4 oz.

Alcohol .... 4 qts.

ESSENCE OF CITRONELLA.

Oil of citronella........................... 3 oz.

Alcohol .... 1 gal.

ESSENCE OF LEMON GRASS (EXTRAIT DE SCHOENANTHE).

Oil of lemon grass......................... 2 oz.

Alcohol .... 1 gal.

EXTRACT OF LILAC (EXTRAIT DE LILAS).

The genuine is seldom made; the preparation sold under this name consists of:

Lilacine .... 2 oz.

Extract of orange flowers, from pomade...... 2 qts.

Extract of tuberose, from pomade........... 3 qts.

Tincture of civet......................... 1 oz.

Oil of Ylang-ylang........................ 2 drams.

Jacinthe (Naef) .... 1 dram.

EXTRACT OF HONEYSUCKLE (EXTRAIT DE CHÈVRE-FEUILLE).

The author has made this extract by treating the pomade prepared from the flowers of Lonicera caprifolium, in the following proportion:

Honeysuckle pomade .... 6 lbs.

Alcohol .... 5 qts.

The commercial extract of this name is always a compound which may be prepared according to the following formula:

Extract of rose, made from the pomade.......	1 qt.
Extract of tuberose, from pomade...........	1 qt.
Extract of orange, from pomade.............	1 qt.
Extract of jasmine, from the pomade........	1 qt.
Oil of neroli..............................	8 grains.
Oil of honeysuckle, syn. (Heine)...........	2 drams.
Oil of jasmine, syn. (Schimmel)............	2 drams.
Musk xylene, 100%........................	1 dram.
Coumarin	1 dram.

ESSENCE OF GERANIUM.

Oil of geranium (rose-geranium)...........	5½ oz.
Alcohol	5 qts.

EXTRACT OF CUCUMBER (EXTRAIT DE CONCOMBRES).

Cucumbers	8 lbs.
Alcohol	5 qts.

The cucumbers are peeled, cut into thin slices, and macerated in the warm alcohol. If the odor is not strong enough in the alcohol after some days, it is poured over some more fresh slices, the macerated residue is expressed, and at the end of the operation all the liquids are united and filtered.

EXTRACT OF HELIOTROPE (EXTRAIT D'HÈLIOTROPE).

Heliotrope pomade	6 lbs.
Alcohol	5 qts.

This has been manufactured only very experimentally; the great majority of the so-called extracts of heliotrope are compounded from:

Extract of rose, from pomade..............	2 qts.
Extract of orange flowers, from pomade......	1 qt.
Tincture of ambergris.....................	4 oz.
Heliotropin	4 oz.
Musk xylene, 100%......................	1 dram.
Alcohol	1 qt.

More recently, piperonal, under the name heliotropin, is used for making this extract—

Heliotropin	½ oz.
Alcohol	1 pint.

It is necessary to blend this with various other aromatics in order to cover the pronounced odor. A little coumarin is usually of great help.

EXTRACT OF JASMINE (EXTRAIT DE JASMIN).

Jasmine pomade	4 lbs.
Alcohol	1 gal.

ESSENCE OF LAVENDER (EXTRAIT DE LAVANDE).

Oil of lavender	7 oz.
Alcohol	5 qts.

A far superior essence may be prepared by the distillation of:

Oil of lavender	7 oz.
Rose water	2 qts.
Alcohol	10 qts.

The distillation is continued until one-half of the entire liquid has passed over; the residue in the still furnishes an essence of lavender of the second quality.

EXTRACT OF WALLFLOWER (EXTRAIT DE GIROFLÉ).

Extract of jasmine, from pomade	1 qt.
Extract of orange flower, from pomade	1 qt.
Extract of rose, from pomade	1 qt.
Tincture of vanilla	1 pint.
Tincture of orris root	1 pint.
Œillet (Naef)	1 oz.
Musk xylene, 100%	1 dram.

EXTRACT OF LILY (EXTRAIT DE LYS).

As to this delightful odor the remark made under the preceding head applies likewise:

Extract of jonquil, from pomade.............	3 pints.
Extract of jasmine, from pomade............	13½ fl. oz.
Extract of orange flower, from pomade.......	27 fl. oz.
Extract of rose, from pomade...............	3 pints.
Extract of tuberose, from pomade...........	3 qts.
Oil of Ylang-ylang........................	2 drams.
Oil of bois de rose........................	2 drams.
Lily syn. (Muhlenthaler)....................	2 oz.
Aurantiol (Givaudan)	1 dram.

EXTRACT OF MAGNOLIA (EXTRAIT DE MAGNOLIA).

This favorite perfume is a mixture of:

Extract of orange flower, from pomade........	2 qts.
Extract of rose, from pomade...............	4 qts.
Extract of tuberose, from pomade...........	1 qt.
Extract of violet, from pomade.............	1 qt.
Oil of citronella java......................	1 dram.
Floressence jasmine	1 oz.
Cyclosia (Naef)	1 oz.
Heliotropin	1 oz.

TINCTURE OF VANILLA.

Vanilla beans, Mexican....................	2 lbs.
Alcohol	1 gal.

TINCTURE OF MUSK (EXTRAIT DE MUSC).

Musk	2 oz.
Alcohol	1 gal.

This tincture is of special importance on account of its useful property of fixing other very volatile odors as well as on account of its own odor which when very dilute is attractive to many people.

EXTRACT OF MYRTLE (EXTRAIT DE MYRTE).

Owing to the small yield of essential oil furnished on distillation by the myrtle flowers and the comparatively high price of the oil of myrtle, nearly all the extract of myrtle is prepared artificially, as follows:

Extract of jasmine, from pomade............	½ pint.
Extract of orange flower, from pomade........	1 qt.
Extract of rose, from pomade................	1 qt.
Extract of tuberose, from pomade..........	1 qt.
Tincture of vanilla........................	1 qt.
Oil myrtle	2 drams.

EXTRACT OF NARCISSUS.

In perfumery, two extracts of narcissus are distinguished— true extract of narcissus, from the flowers of the garden plant, Narcissus poeticus, and the so-called extract of jonquille, from Narcissus Jonquilla, which is cultivated in Southern France and whose odor is obtained by maceration. Genuine extract of narcissus is even more rarely obtainable than extract of jonquille; the odors of both are imitated, mainly according to the following compositions:

1. EXTRACT OF NARCISSUS (EXTRAIT DE NARCISSE).

Extract of jonquille, from pomade...........	2 qts.
Extract of tuberose, from pomade...........	2 qts.
Tincture of storax........................	1 oz.
Narcissus, syn. (Verley)...................	3 oz.

2. EXTRACT OF JONQUILLE (EXTRAIT DE JONQUILLE).

Extract of jonquille, from pomade...........	3 qts.
Extract of orange flower, from pomade.......	1 qt.
Extract of tuberose, from pomade...........	1 qt.
Tincture of vanilla........................	½ pint.

ESSENCE OF CLOVE (EXTRAIT DE CLOUS DE GIROFLES).

Oil of clove..............................	4½ oz.
Alcohol	5 qts.

EXTRACT OF PINK (EXTRAIT D'ŒILLET).

Floressence rose	3 drams.
Floressence jasmine	2 drams.
Floressence tuberose	1 dram.
Tincture of orris root	16 oz.
Œillet (Naef)	1 oz.
Heliotropol	2 drams.
Jacinthe (De Laire)	2 drams.
Oil Ylang-ylang, Manilla	2 drams.
Cyclosia	1 dram.
Musk xylene, 100%	1 dram.
Alcohol, to make	4 qts.

EXTRACT OF ORANGE FLOWER OR NEROLI (EXTRAIT DE FLEURS D'ORANGES, EXTRAIT DE NÉROLI).

Orange-flower pomade	4 lbs.
Alcohol	1 gal.

Or,

Oil neroli pétale	2 oz.
Alcohol	1 gal.

EXTRACT OF PATCHOULY.

Extract of orange flowers, from the pomade	1 qt.
Extract of rose, from the pomade	2 qts.
Extract of jasmine, from the pomade	1 qt.
Tincture of musk Tonquin	3 oz.
Tincture of civet	2 oz.
Tincture of oakmoss	8 oz.
Oil of patchouly	1 oz.
Otto of rose, Bulgarian	2 drams.
Oil of sandalwood	2 drams.
Coumarin	2 drams.
Vanillin	1 dram.
Musk xylene, 100%	1 dram.

TINCTURE OF BALSAM OF PERU (EXTRAIT DE PEROU).

Peru balsam .... 8 oz.
Alcohol .... 1 gal.

This tincture, though of a very pleasant odor, can be used only for scenting soaps or sachets, as it has a very dark brown color; by distilling alcohol over Peru balsam a colorless extract is obtained, though of a fainter odor.

EXTRACT OF SWEET PEA (EXTRAIT DE POIS DE SENTEUR).

This extract, made only occasionally in Southern France by maceration of the pomade, is but rarely met with in commerce; what passes under this name is made as follows:

Extract of orange flower, from pomade....... 2½ pints.
Extract of rose, from pomade.............. 2½ pints.
Extract of tuberose, from pomade........... 2½ pints.
Oil of sweet pea, syn....................... 2 oz.

EXTRACT OF RESEDA (EXTRAIT DE MIGNONETTE).

Extract of reseda, from the pomade......... 1½ qts.
Extract of rose, from the pomade........... 1 qt.
Extract of violet, from the pomade.......... 1 qt.
Tincture of orris root............. .... 1 pint.
Mignonette, syn. (Nadel)................... 1 oz.
Cyclosia .... 1 dram.
Jasmine, syn. (Naarden)................... 1·dram.
Musk xylene, 100%...................... 1 dram.

ESSENCE OF EXTRACT OF ROSE (EXTRAITS DE ROSE).

In commerce several sorts of essence or extract of rose are distinguished; only the cheaper grades are made by direct solution of the oil of rose in alcohol, the better grades are prepared only from pomades. As the rose is the noblest of flow-

ers, so are these odors the most magnificent thus far produced
by the art of perfumery, since they are approached in delicacy
and fragrance only by the genuine extracts of orange flower
and violet.   The so-called rose waters (eaux de rose) are best
obtained by distillation of fresh or salted rose leaves with water.
The preceding formulæ will show that both extract of rose
and rose water form important constituents of many compound
essences, hence these materials require special attention.   In
the following pages we enumerate only those formulæ which
are acknowledged as the best and furnish the finest product.
As rose water likewise belongs among the rose odors we give
directions for its preparation, and observe in passing that the
precautions required in the manufacture of this one apply
also to all aromatic waters (eaux aromatisées).   The first
essential to the production of a fine aromatic water is the employment of the freshest possible flowers; when kept in stock,
chemical changes occur in the leaves which affect also the
aromatic constituents and lead to a deterioration of the fragrance.   Hence we urgently recommend to distil the freshly
gathered flowers as soon as possible, even if the quantity on
hand be small.   Should this not be feasible, it is advisable to
press the flowers immediately after gathering in stone-ware
pots and to pour over them a saturated solution of table salt.
A concentrated saline solution prevents decomposition by
the abstraction of water; and thus larger quantities of flowers may be gathered and distilled with the salt solution.   The
majority of aromatic waters are prepared in this way, for instance, rose, jasmine, lilac, and others.   They enter less into
handkerchief perfumes than into various mouth and other
washes, and cosmetics in general.

ROSE WATER (EAU DE ROSE TRIPLE).

Rose leaves.......................... 4 lb.
Water.............................. 20 pints.
Mix them, and by means of steam, distil 10 pints.

The average perfumer will find it much to his advantage to purchase his rose-water rather than to prepare it. It is obtained as a by-product in the preparation of otto of rose and if purchased from reliable sources will be found satisfactory. Strength and quality are more important than the price.

ESSENCE OF [OIL OF] ROSE (ESPRIT DE ROSES TRIPLE).

Oil of rose.............................	3 oz.
Alcohol	1 gal.

This will be found to be cheaper than the extract.

EXTRACT OF CHINA ROSES (ESSENCE DE ROSES JAUNES).

Essence of rose (triple)...................	2 qts.
Tincture of tonka........................	½ pint.
Extract of jasmine.......................	2 qts.
Oil of rose, syn. (Bush)..................	2 oz.
Musk xylene, 100%......................	1 dram.

EXTRACT OF SWEET-BRIER (WILD ROSE) (EXTRAIT D'EGLANTINE).

Extract of cassie, from pomade.............	44 fl. oz.
Extract of orange flower, from pomade.......	44 fl. oz.
Extract of rose, from pomade..............	2½ qts.
Essence of rose (triple)...................	44 fl. oz.
Oil of neroli............................	¼ oz.
Rose rouge, syn. (Naef)..................	¼ oz.

EXTRACT OF MOSS-ROSE (EXTRAIT DE ROSES MOUSSEUSES).

Extract of rose, from pomade..............	2 qts.
Extract of orange flower, from pomade.......	1 qt.
Essence of rose (triple)...................	1 qt.
Tincture of ambergris....................	1 pint.
Tincture of musk........................	4 oz.

Extract of Tea-Rose (Extrait de Rosa théa).

Extract of rose, from pomade...............	1 qt.
Extract of jasmine, from pomade...........	1 qt.
Extract of orange flower, from pomade........	½ pint
Essence of rose (triple)....................	1 qt.
Tincture of musk.........................	4 oz.
Tincture of orris root.....................	½ pint.

Extract of White Rose (Essence de Roses blanches).

Extract of rose, from pomade................	2 qts.
Extract of jasmine, from pomade...........	2 qts.
Extract of cassie, from pomade.............	1 qt.
Oil of patchouly..........................	1 dram.
Otto of rose..............................	5 drams.
Musk xylene, 100%.......................	1 dram.

Extract of Twin-Roses (Essences de Roses jumelles).

Extract of rose, from pomade...............	5 qts.
Oil of rose................................	1¾ oz.

Extract of Santal (Extrait de Santal).

Tincture of santal........................	3½ oz.
Essence of rose (triple)...................	1 pint.
Alcohol	9 pints.

Tincture of Galbanum.

Gum galbanum	2 lbs.
Alcohol	4 qts.

Tincture of Tolu (Extrait de Baume de Tolou).

Tolu balsam	2 lbs.
Alcohol	1 gal.

The remark made under tincture of storax applies also to this.

Tincture of Tonka (Extrait de Tonka).

Tonka beans, crushed......................	2 lbs.
Alcohol	1 gal.

Extract of Tuberose (Extrait de Tuberose).

Tuberose pomade	4 lbs.
Alcohol	1 gal.

Tincture of Costus.

Costus	2 lbs.
Alcohol	4 qts.

Extract of Violet (Extrait de Violette).

Violet pomade	4 lbs.
Alcohol	1 gal.

This extract is very expensive; a good imitation is made as follows:

Extract of violet, from pomade.............	1 qt.
Extract of cassie, from pomade.............	1 pint.
Extract of rose, from pomade..............	1 pint.
Extract of jasmine, from pomade............	1 pint.
Tincture of orris.........................	1 pint.
Ionone (De Laire).......................	4 drams.
Violae (Vidal-Charvet)...................	2 drams.
Heliotropin	2 drams.
Oil Ylang-ylang, Manila..................	1 dram.
Musk xylene, 100%.....................	1 dram.

This is a very superior violet perfume, which is far cheaper than the true extract of violet, a product which seldom or never is sold as such on account of the prohibitive cost.

EXTRACT OF VERBENA (EXTRAIT DE VERVEINE).

True oil of verbena is rather expensive. Hence artificial compositions are employed under the name of verbena which resemble the true odor, though not exactly like it.

EXTRACT OF VERBENA A.

Citral	75 grains.
Oil of lemon.............................	14 oz.
Oil of orange	3½ oz.
Alcohol	5 qts.

This extract is cheap and is used immediately as a perfume. The extract usually sold under the French name Extrait de verveine is more expensive and far superior:

EXTRACT OF VERBENA B.

Extract of orange flower, from pomade.......	30 fl. oz.
Extract of rose, from pomade...............	1 qt.
Extract of tuberose, from pomade...........	¹/₃ oz.
Oil of citron zeste........................	½ oz.
Citral	¾ oz.
Oil of lemon.............................	9 oz.
Oil of orange	4½ oz.
Alcohol	4²/₃ pints.

EXTRACT OF VOLCAMERIA (EXTRAIT DE VOLCAMERIA).

This extract is no more derived from the fragrant blossom whose name it bears than are those of the lily, pink, and others met with in commerce. It is prepared according to the following formula:

Extract of jasmine, from pomade............	1 pint.
Extract of rose, from pomade...............	1 qt.
Extract of tuberose, from pomade...........	2 qts.
Extract of violet, from pomade.............	2 qts.
Tincture of musk.........................	½ pint.

Extract of Gardenia.

Floressence jasmine	4 drams.
Floressence tuberose	2 drams.
Floressence orange flowers	2 drams.
Oil jasmine, syn. (Naef)	2 drams.
Oil neroli petale	1 dram.
Oil bergamot	4 drams.
Cyclosia	1 dram.
Narcisse (Givaudan)	1 dram.
Imex (Vidal-Charvet)	1 dram.
Musk xylene, 100%	1 dram.
Alcohol, to make	4 qts.

Extract of Wintergreen (Extrait de Gaulthérie).

This essence is more commonly sold under the English than the French name. Its composition is the following:

Tincture of ambergris	1 pint.
Extract of cassie	1 qt.
Essence of lavender	1 pint.
Extract of orange flower, from pomade	1 qt.
Extract of rose, from pomade	2 qts.
Tincture of vanilla	1 pint.
Essence of vetiver	1 pint.

It must be borne in mind by the reader that while the tinctures and simple extracts mentioned in this chapter are suitable only for use as raw materials in building up harmonious, finished odors others are already composed of a large number of oils and are practically completed perfumes suitable for sale as such. This does not mean that they are not sometimes used to advantage as constituents of other perfumes and it is for this reason that they have been included here. The extracts and toilet waters mentioned in subsequent chapters are in almost all cases finished perfumes and must be considered as such and the exercise of some discretion will be required on the part of the reader to decide whether the term extract is being applied to a semifinished material or to a completed perfume.

CHAPTER XIV.

THE DIVISION OF PERFUMERY.

ACCORDING to the purposes for which they are intended, the various articles of perfumery may be divided into several groups, which include not only the true perfumes which are sold and used as such but also the products in which the perfume ingredients are incidental but scarcely less important on that account. They are:

TRUE PERFUMES.

A. *Liquid.*—Alcoholic handkerchief perfumes. Among these are the so-called extracts, bouquets, and waters. Ammoniacal and acid perfumes: aromatic vinegars and volatile ammoniacal salts.

B. *Dry.*—Sachet powders, fumigating pastils and powders.

PREPARATIONS FOR THE CARE OF THE SKIN.

Emulsions, crêmes, perfumed soaps, toilet waters, nail powders.

PREPARATIONS FOR THE CARE OF THE HAIR.

Hair oils, pomades, hair washes, brilliantines.

PREPARATIONS FOR THE CARE OF THE MOUTH.

Tooth powders, mouth washes.

COSMETICS.

Paints, powders, hair dyes, depilatories, etc.

In connection with the description of these different articles some remarks will be made about the colors employed in perfumery and about the utensils used with the cosmetics, such as combs, brushes, sponges, etc.

CHAPTER XV.

THE MANUFACTURE OF HANDKERCHIEF PER-FUMES, BOUQUETS, OR AROMATIC WATERS.

THE manufacture of handkerchief perfumes is very sim-ple: the extracts prepared as directed in Chapter XIII. are mixed in definite proportions and the perfume is finished. If the extracts are well seasoned, the perfumes blend in perfect harmony within a few days, and this time may be even short-ened by the use of the apparatus illustrated in Fig. 32. If the extracts have been but recently prepared, a longer time will be required before the odor of the alcohol and the seve-ral constituents is imperceptible and all odors have blended into a harmonious whole.

If the manufacturer can afford to allow the finished ex-tracts and perfumes to season for some length of time—of course, in well-closed and completely filled vessels—in a cool place, they will improve markedly in quality. Perfumes which contain but a single odor or in which a certain odor distinctly predominates are usually called by the name of the respective plant, etc., under a French title, *e. g.*, extrait de violette, extrait de reséda, etc. Combinations of many odors which produce an agreeable impression as a whole, while no one odor predom-inates, are called bouquets or waters; for instance, Bouquet de Jockey Club, Eau de Mille Fleurs, Cologne Water, Hun-garian Water, etc.

The mixture of the extracts is effected in strong glass bot-tles of a capacity exactly adapted to the perfume, so as to be completely filled. For perfumes which require seasoning to make the odors blend we use small glass balls of which enough are introduced into the bottle to make the mixture rise into the neck of the container which is then closed air-tight and preserved in a dark, cool place.

Of course, all perfumes should be perfectly clear and free from turbidity. The extracts made from pomades or essential oils are clear and furnish perfumes that remain so; extracts prepared from balsams or resins should be allowed to stand at rest for several weeks and then be carefully decanted from the sediment. Filtration should be dispensed with unless absolutely unavoidable, on account of the large amount of oxygen with which the extract would thereby come in contact, to the detriment of the odor.

The bottles in which the perfumes are mixed, as well as those in which they are put up for sale, must be perfectly dry, as a very small amount of water often suffices to separate a portion of the aromatics and to render the liquid turbid or opalescent.

Fine perfumes are always sold in glass vessels with ground-glass stoppers; cork has a peculiar odor which it would communicate to the liquid. For the more perfect exclusion of the air the stoppers and bottle necks are moreover covered with animal membrane, sheet rubber, or vegetable parchment, with an outer cap of white glove leather.

In the case of very expensive perfumes, much care is bestowed on the container; certain perfumes are filled into bottles of peculiar form and color, or into small porcelain jars provided with corresponding labels printed in gold and colors. Sometimes the container costs many times the price of the perfume. But as the finest perfumes are articles of luxury in the truest sense of the word, they require extreme care in their putting up; and good taste in the selection of the containers for fluids, pomades, cosmetics, powders, etc., is of as much importance to the perfumer as the possession of a sensitive and trained olfactory organ.

In the following formulas for the preparation of bouquets, the words extract, essence, and tincture have the same meaning, as was explained under Chapter XIII.

In many cases the cost of these preparations may be reduced and their manufacture simplified by the use of synthetic oils, and accordingly a number of such formulas are given as examples of modern practice.

CHAPTER XVI.
FORMULAS FOR HANDKERCHIEF PERFUMES.
BOUQUET DE L'ALHAMBRA.

Extract of rose......................	1 pint.
Extract of orange flower..............	1 pint.
Musk ambrene, 10%.................	1 dram.
Extract of tuberose..................	2 qts.
Tincture of civet....................	4 oz.
Resinarome orris	1 oz.

EXTRAIT D'AMBRE, I.

Tincture of ambergris.................	8 oz.
Tincture of musk....................	8 oz.
Oil of rose..........................	1 oz.
Grisambrene	4 drams.
Alcohol	4 qts.

EXTRAIT D'AMBRE, II.

Essence of rose (triple)...............	2 qts.
Tincture of ambergris	4 qts.
Tincture of musk....................	1 qt.
Tincture of vanilla...................	1 pint.

BOUQUET DE L'AMOUR.

Extract of cassie....................	1 qt.
Tincture of ambergris................	4 oz.
Extract of jasmine...................	1 qt.
Tincture of musk....................	5 oz.
Extract of violet....................	1 qt.
Extract of rose.....................	1 qt.

Baisers du Printemps (Spring Kisses).

Ariel	1 oz.
Tincture of ambergris	3 fl. oz.
Extract of jasmine	6 fl. oz.
Extract of rose	5 pints.
Extract of violet	5 pints.
Essence of rose (triple)	10 fl. oz.
Oil of bergamot	120 grains.
Oil of lemon	30 grains.
Iralia	2 drams.
Coumarin	1 dram.
Musk xylene, 100%	1 dram.

Eau de Berlin.

Oil of anise	150 grains.
Oil of bergamot	1 oz.
Oil of cardamom	15 grains.
Oil of lemon	30 grains.
Oil of coriander	15 grains.
Oil of geranium	30 grains.
Oil of melissa	15 grains.
Oil of neroli	75 grains.
Oil of rose	30 grains.
Oil of santal	30 grains.
Oil of thyme	15 grains.
Alcohol	10 qts.

Buckingham Flowers.

Extract of cassie	1 qt.
Tincture of ambergris	4 oz.
Extract of jasmine	1 qt.
Extract of orange flower	1 qt.
Extract of rose	1 qt.
Tincture of orris root	1 pint.
Oil of lavender	40 grains.
Oil of neroli	40 grains.
Oil of rose	75 grains.

Bouquet Adoré.

Extract of jasmine.................... 1 pint.
Extract of rose....................... 1 pint.
Extract of tuberose................... 1 pint.
Extract of violet..................... 1 pint.
Tincture of orris root................ 1 pint.
Oil rose, syn. (Chiris)............... 75 grains.

Bouquet du Bosphore.

Extract of cassie..................... 1 qt.
Extract of jasmine.................... ½ pint.
Extract of tuberose .... ½ pint.
Tincture of civet..................... 18 grains.
Essence of rose (triple) .... ½ pint.

Bouquet des Chasseurs.

Extract of cassie..................... 20 fl. oz.
Tincture of musk .... 10 fl. oz.
Extract of neroli..................... 20 fl. oz.
Extract of orange flower.............. 20 fl. oz.
Tincture of tonka bean .... 40 fl. oz.
Tincture of orris root................ 20 fl. oz.
Oil of lemon.......................... ½ oz.
Essence of rose (triple) .... 5 pints.

Bouquet de la Cour.

Tincture of ambergris .... 2 oz.
Extract of jasmine.................... 1 qt.
Tincture of musk...................... 2 oz.
Extract of rose....................... 1 qt.
Extract of violet .... 1 qt.
Essence of rose (triple).............. 1 qt.
Oil of bergamot .... 45 grains.
Oil of lemon .... 45 grains.
Oil of neroli .... 45 grains.

BOUQUET DE CHYPRE.

Musk ambrene, 100%.....................	1 dram.
Tincture of musk.........................	8 oz.
Tincture of tonka........................	1 qt.
Tincture of vanilla.......................	1 qt.
Tincture of orris root.................⸴.....	1 qt.
Essence of rose (triple)....................	2 qts.
Tincture oak moss........................	8 oz.
Resinoid labdanum (Rourc)...............	1 oz.
Coumarin	4 drams.
Alcohol, to make........................	8 qts.

BOUQUET DES DÉLICES.

Tincture of ambergris.....................	4 oz.
Extract of rose..........................	1 qt.
Extract of tuberose.......................	1 qt.
Extract of violet.........................	1 qt.
Tincture of orris root.....................	1 pint.
Oil of bergamot..........................	½ oz.
Dianthine	1 oz.

BOUQUET DE FLEURS (NOSEGAY).

Tincture of benzoin.......................	4 oz.
Extract of rose...........................	3 pints.
Extract of tuberose.......................	3 pints.
Extract of jasmine........................	3 pints.
Oil of bergamot..........................	2 oz.
Neroli, syn., Morana......................	1 oz.
Musk xylene, 100%.......................	1 dram.
Alcohol	1 qt.

CONVALLERIA (LILY OF THE VALLEY, FLEURS DE MAI).

Extract of tuberose.......................	1½ pints.
Extract of jasmine........................	1½ pints.
Extract of orange flower...................	1½ pints.
Extract of rose...........................	1½ pints.
Oil of bois de rose........................	2 drams.
Oil of Ylang-ylang........................	2 drams.
Valley lily, syn. (Polak and Swartz).........	1 oz.

COURONNE DE FLEURS (GARLAND OF FLOWERS).

Extract of cassie........................... 20 oz.
Tincture of ambergris...................... 10 oz.
Extract of jasmine......................... 20 oz.
Tincture of musk.......................... 10 oz.
Tincture of orris root.................. 5 pints.
Oil of bergamot........................ 1½ oz.
Oil of lavender........................ 1½ oz.
Oil of clove .... 75 grains.
Oil of neroli........................... 1½ oz.
Oil of rose .... 1½ oz.
Alcohol .... 5 pints.

COURT BOUQUET.

Oil of bergamot........................ ⅜ oz.
Oil of neroli........................... 24 grains.
Alcohol................................ 5½ oz.
Orris root............................. 1 oz.
Storax, liquid......................... 8 grains.
Musk.................................. 3 grains.

Macerate for two weeks, and filter.

ESTERHAZY BOUQUETS.

An old renowned perfume, a former rival of Cologne water; the name is derived from a noble Hungarian family.

A. BOUQUET D'ESTERHAZY (FRENCH FORMULA).

Tincture of ambergris.................. ½ pint.
Extract of neroli...................... 1 qt.
Extract of orange flower............... 1 qt.
Tincture of tonka...................... 1 qt.
Tincture of vanilla .... 1 qt.
Tincture of vetiver.................... 1 qt.
Tincture of orris root................. 1 qt.
Essence of rose (triple) .... 1 qt.
Oil of clove .... 75 grains.
Oil of santal......................... 75 grains.

B. Bouquet Esterhazy (German Formula).

Calamus root .... 3 oz.
Cloves .... 3 oz.
Nutmeg .... 3 oz.
Alcohol .... 4 qts.

Macerate for two weeks and filter; in the filtrate dissolve:

Tincture of ambergris..................... 6 oz.
Ammonia .... 30 grains.
Oil of bitter almond...................... 30 grains.
Oil of lemon............................. 3 oz.
Tincture of musk......................... 6 oz.
Oil of neroli............................ 60 grains.
Oil of orange peel........................ 30 grains.
Oil of rose......................,....... 75 grains.

Bouquet Lebanon.

Essence of rose, absolute (Roure)............ 3 drams.
Essence of jasmine, absolute (Roure)........ 2 drams.
Eliator (Vidal-Charvet) .... 4 oz.
Musk xylene .... 1 dram.
Alcohol :................................ 4 qts.

Fiori d'Italia

Extract of cassie......................... 1 pint.
Tincture of ambergris..................... 5 oz.
Extract of jasmine........................ 1 qt.
Tincture of musk......................... 5 oz.
Extract of rose........................... 2 qts.
Extract of violet......................... 1 qt.
Essence of rose (triple).................. 1 qt.

Lilac (Extrait de Lilas).

Lilacine .... 3 oz.
Extract of jasmine........................ 2 qts.
Extract of tuberose....................... 3 qts.
Jacinthe (Heine) .... 4 drams.
Cyclosia .... 2 drams.

ESSENCE DES BOUQUETS, A (ESS. BOUQUET).

Tincture of ambergris....................	4 oz.
Tincture of orris root.....................	2 qts.
Essence of rose (triple)....................	2 qts.
Oil of bergamot..........................	1 oz.
Oil of lemon.............................	½ oz.

ESS. BOUQUET, B.

Extract of rose, from pomade...............	3 pints.
Extract of jasmine, from pomade............	3 pints.
Tincture of orris........................	1 pint.
Tincture of vanilla.......................	8 oz.
Tincture of Musk Tonkin..................	5 oz.
Tincture of tonka........................	4 oz.
Tincture of civet........................	3 oz.
Tincture of ambergris.....................	2 oz.
Oil of bergamot..........................	1 oz.
Otto of rose.............................	1 oz.
Oil of neroli pétale......................	1 dram.
Oil of mace.............................	1 dram.
Oil of coriander.........................	1 dram.

This perfume is much admired in England. The title Ess. Bouquet is an abbreviation of the full name given above.

ESS. BOUQUET, C.

Extract of rose, from pomade...............	2½ pints.
Extract of jasmine, from pomade............	2½ pints.
Tincture of orris........................	2½ pints.
Oil of bergamot..........................	1 oz.
Oil of rose..............................	¾ oz.

FLORIDA.

Floressence orange flowers.................	6 drams.
Floressence jasmine......................	2 drams.
Oil bergamot, terpeneless.................	2 drams.
Aurantiol (Givaudan)	4 drams.
Alcohol, to make.........................	4 qts.

Bouquet de Flore.

Extract of rose..........................	1 qt.
Extract of orange flower...................	1 pint.
Extract of tuberose........................	1 pint.
Extract of violet..........................	½ pint.
Tincture of benzoin.......................	3 fl. oz.
Tincture of storax.......................	3 fl. oz.
Tincture of musk.......................	1½ fl. oz.
Oil of citronella..........................	¾ oz.
Alcohol	2 qts.

Honeysuckle (Extrait de Chèvre-feuille).

Honeysuckle, syn. (Polak).................	4 oz.
Extract of orange flowers..................	2 qts.
Extract of jasmine........................	1 qt.
Extract of tuberose.......................	1 qt.
Musk xylene, 100%.....................	1 dram.

Heliotrope (Extrait d'Héliotrope).

Extract of jasmine, from pomade............	2 qts.
Extract of orange flowers..................	1 pint.
Tincture of ambergris.....................	4 oz.
Heliotropine	4 oz.
Ylang Ylang, Siegert......................	2 drams.
Coumarin	2 drams.
Musk xylene	1 dram.
Alcohol, to make.......	4 qts.

White Heliotrope.

Floressence jasmine, colorless..............	1 oz.
Oil neroli pétale..........................	1 dram.
Ylang Ylang, Siegert......................	1 dram.
Heliotropol	4 oz.
Cyclosia	2 drams.
Ionone (de Laire).........................	1 dram.
Alcohol	4 qts.

New-Mown Hay.

Hay owes its fragrance partly to cumarin, which is present in many plants, but in especially large amount in tonka beans. Hence all similar perfumes must contain tincture of tonka. Other aromatic substances, however, contribute to the odor of hay, but the cumarin gives, as it were, the key-note to its real odor.

A very pleasant perfume is made after the following formula:

Essence of rose (triple)	1 qt.
Coumarin	1 oz.
Extract of jasmine	1 qt.
Extract of orange flower............	1 qt.
Extract of rose	1 qt.
Tincture of tonka....................	2 qts.

Some add to this perfume 1 pint of extract of cassie which imparts a greenish color to it.

Royal Horse-Guard's Bouquet.

Extract of orange flower	20 fl. oz.
Tincture of musk....................	10 fl. oz.
Extract of rose......................	5 pints.
Tincture of vanilla	20 fl. oz.
Tincture of orris root	20 fl. oz.
Oil of clove	120 grains.

Bouquet d'Irlande.

Extract of white rose.................	5 qts.
Tincture of vanilla...................	1 lb.

An exceedingly fine perfume.

Hovenia.

This plant, Hovenia dulcis, indigenous to Japan, has a peculiar odor, which, however, is not pleasant to European taste. The perfume sold under this name has a special odor,

though it differs from that of the plant.   It is made according
to the following formula:

Oil of lemon...........................	3 oz.
Oil of clove...........................	¼ oz.
Oil of neroli...........................	75 grains.
Oil of rose	75 grains.
Alcohol...............................	5 qts.

HUNTSMAN'S NOSEGAY.

Essence of rose (triple)................	1 pint.
Extract of cassie......................	6 fl. oz.
Extract of orange flower...............	6 fl. oz.
Tincture of musk......................	150 grains.
Tincture of tonka.....................	1 pint.
Oil of citronella......................	150 grains.
Alcohol...............................	3 qts.

BOUQUET DU JAPON.

Extract of rose.......................	1 qt.
Extract of orange flower...............	1 qt.
Essence of patchouly...................	½ pint.
Extract of verbena....................	1 pint.
Essence of vetiver....................	1 pint.
Tincture of civet.....................	3 fl. oz.
Tincture of musk......................	⅓ fl. oz.

EAU JAPONAISE.

Tincture of orris root	1 qt.
Essence of patchouly	1 qt.
Extract of santal.....................	1 qt.
Extract of verbena	1 qt.
Essence of vetiver....................	1 pint.
Essence of rose (triple)	1 qt.

JOCKEY CLUB.

France first introduced a perfume under this name, which
soon became popular and was largely imitated. Jockey Club
perfume is among the finest known to the trade; the delicacy

f its odor rests largely on the extracts of rose and jasmine
which are employed in their strongest form—an alcoholic extract of a pomade well charged with the odors of the flowers.
As in the case of Cologne water, there are a number of widely
diverging formulas for its preparation, from which we select
few which furnish excellent perfumes.

JOCKEY CLUB, A (ENGLISH FORMULA).

Extract of jasmine............................	1 pint.
Tincture of ambergris.........................	¾ pint.
Extract of rose...............................	1½ pints.
Extract of tuberose...........................	¾ pint.
Tincture of orris root.........................	3 pints.
Essence of rose (triple).......................	1½ pints.
Oil of bergamot..............................	¾ oz.

JOCKEY CLUB, B (FRENCH FORMULA).

Oil of bergamot..............................	1 oz.
Extract of jasmine............................	3 pints.
Extract of rose...............................	2 qts.
Extract of tuberose...........................	1 qt.
Tincture of civet.............................	½ pint.
Oil of mace..................................	1 oz.

JOCKEY CLUB, C (GERMAN FORMULA).

Tincture of ambergris.........................	5 fl. oz.
Extract of jasmine......	1 qt.
Extract of rose...............................	3 pints.
Extract of tuberose...........................	1 qt.
Extract of violet.............................	1 pint.
Tincture of civet.............................	5 fl. oz.
Oil of bergamot..............................	¾ oz.
Oil of neroli.................................	½ oz.

JONQUILLE (EXTRAIT DE JONQUILLE).

Extract of jonquil............................	2 qts.
Extract of orange flower......................	1 qt.
Narcisse, syn. (Naarden).....................	4 drams.
Tincture of vanilla...........................	2 oz.

KISS ME QUICK.

Extract of cassie......................	1 qt.
Extract of ambergris....................	½ pint.
Extract of narcissus (Jonquille)..........	2 qts.
Tincture of tonka......................	1 qt.
Tincture of orris root..................	2 qts.
Tincture of civet......................	½ pint.
Essence of rose (triple)................	1 qt.
Oil of citronella......................	75 grains.
Oil of lemon grass.....................	45 grains.

This perfume, which was once very popular, owes its peculiar refreshing odor to the tincture of tonka beans; by increasing this ingredient the specific odor can be made more pronounced.

BOUQUET COSMOPOLITE.

Extract of jasmine....................	1 pint.
Essence of lavender...................	½ pint.
Tincture of musk.....................	½ pint.
Essence of patchouly.................	½ pint.
Extract of santal....................	½ pint.
Extract of tuberose..................	1 pint.
Tincture of vanilla..................	½ pint.
Extract of violet....................	1 qt.
Essence of rose (triple).............	1 pint.
Coumarin	2 drams.
Musk xylene, 100%...................	1 dram.

COLOGNE WATER (EAU DE COLOGNE).

This famous perfume, which was first made in Cologne on the Rhine, its formula being kept secret, can be produced anywhere of the same quality as the original. In order to obtain a first-class product, it is necessary, besides using the finest oils—a matter of course for all fine perfumes—to observe another special point. Every Cologne water contains oils of the citron group which develop their best odors only in true spirit of *wine*. Unless an alcohol distilled from *wine* is used, it will be impossible to make a Cologne water of really first quality. While it is possible to make a good cologne with grain or potato spirit, especially if highly rectified, compari-son with one prepared from pure spirit of *wine* will at once show a marked difference. The small amount of œnanthic ether, hardly demonstrable by chemical tests but present in every spirit of wine, exerts a decided influence on the flavor.

Cologne water of the most superior and incomparable qual-ity is made by dissolving the essential oils, excepting the oils of rosemary and neroli, in the alcohol and distilling it, the other oils being added to the distillate.

A very large number of formulas for the preparation of Cologne water have been published of which we subjoin a few. We have purposely omitted those containing many essential oils, as experience has taught us that they are of lit-tle value; for it is not the number of oils that determines the fineness of a perfume, but the manner in which certain odors are combined.

A. Finest Cologne Water (Eau de Cologne Supé-rieure).

Oil of bergamot	3½ oz.
Oil of lemon	5 oz.
Oil of neroli pétale	3½ oz.
Oil of bigarade	1¼ oz.
Oil of rosemary	2½ oz.
Alcohol	30 qts.

B. Cologne Water (Second Quality).

Oil of bergamot	4½ oz.
Oil of lemon	4½ oz.
Oil of neroli pétale	¾ oz.
Oil of orange	4½ oz.
Oil of petit grain	2½ oz.
Oil of rosemary	2½ oz.
Alcohol	30 qts.

C. Cologne Water (ordinary).

Oil of bergamot...................... 7 oz.
Oil of lemon......................... 3½ oz.
Oil of lavender...................... 3½ oz.
Alcohol .... 30 qts.

D. Cologne Water.

Oil of bergamot...................... 1¾ oz.
Oil of lemon......................... 3½ oz.
Oil of lavender......................150 grains.
Oil of neroli........................ ½ oz.
Oil of rosemary...................... 75 grains.
Alcohol .... 30 qts.

E. Cologne Water.

Oil of bergamot...................... 2 oz.
Oil of lemon......................... 1 oz.
Oil of lavender...................... ½ oz.
Oil of melissa....................... ¼ oz.
Oil of neroli........................ ¼ oz.
Alcohol .... 30 qts.

F. Cologne Water.

Oil of bergamot...................... 3½ oz.
Oil of lemon......................... ½ oz.
Oil of lavender...................... ¼ oz.
Oil of melissa....................... ½ oz.
Oil of neroli........................ ¼ oz.
Alcohol .... 30 qts.

G. Cologne Water.

Oil of bergamot...................... 1 lb.
Oil of lemon......................... 1 lb.
Oil of lavender...................... 6½ oz.
Oil of neroli........................ ¾ oz.
Oil of petit grain................... 1½ oz.
Oil of orange .... 1 lb.
Oil of rosemary ....150 grains.
Alcohol .... 30 qts.

H. Cologne Water.

Oil of bergamot........................	2¼ oz.
Oil of cajuput.........................	½ oz.
Oil of lemon...........................	4½ oz.
Oil of lavender........................	6½ oz.
Oil of neroli..........................	2¼ oz.
Oil of orange	4½ oz.
Oil of petit grain.....................	½ oz.
Orange-flower water....................	1 qt.
Alcohol	30 qts.

The numerous formulas show that oils of lemon, bergamot, and orange form normal constituents of every Cologne water; the finer grades always contain, in addition, oils of rosemary and neroli. It is advisable to dissolve the aromatics in very strong alcohol and then to effect the dilution required with orange-flower or rose water. This dilution is also to be employed when a cheaper product is desired.

Lavender Perfumes.

English (Mitcham) oil of lavender should always be used when it is desired to produce perfumes of first quality.

Eau de Lavande Ambrée.

Oil of bergamot.......................	1 oz.
Oil of lemon..........................	½ oz.
Oil of geranium.......................	75 grains.
Oil of lavender (Allen)...............	5½ oz.
Musk xylene, 100%....................	8 grains.
Peru balsam	2 oz.
Storax	4¼ oz.
Civet	15 grains.
Alcohol	10 qts.

The essential oils are dissolved in the alcohol, the other substances are macerated in the solution for one month, and the liquid decanted.

EAU DE LAVANDE DOUBLE.

Tincture of musk......................	3 fl. oz.
Tincture of vanilla....................	3 fl. oz.
Tincture of civet.....................	3 fl. oz.
Oil of bergamot......................	1¼ oz.
Oil of lemon.........................	¾ oz.
Oil of lavender......................	3½ oz.
Rose water (triple)	1 qt.
Alcohol..............................	10 qts.

EAU DE LAVANDE A MILLE FLEURS.

Tincture of ambergris..................	½ pint.
Essence of lavender	2 qts.
Eau de mille fleurs (see below, page 186)..	2 qts.

LEAP-YEAR BOUQUET.

Extract of jasmine	3 pints.
Essence of patchouly..................	1½ pints.
Essence of santal.....................	1½ pints.
Extract of tuberose...................	1 qt.
Extract of verbena...................	6½ fl. oz.
Essence of vetiver....................	1½ pints.
Essence of rose (triple)	1½ pints.

EAU DE LEIPSIC.

Oil of lemon.........................	¾ oz.
Oil of neroli.........................	¾ oz.
Oil of orange	150 grains.
Oil of bergamot.....................	2¼ oz.
Oil of rosemary.....................	75 grains.
Orange-flower water..................	1 qt.
Alcohol..............................	9 pints.

WALLFLOWER (EXTRAIT DE GIROFLÉ).

Extract of jasmine...................	1 qt.
Extract of orange flower.............	1 qt.
Extract of rose......................	1 qt.
Tincture of vanilla...................	1 pint.
Tincture of orris root................	1 pint.
Wall flower, syn. (Morena)...........	2 oz.

LILY (EXTRAIT DE LYS).

Extract of jonquil	3 qts.
Extract of jasmine	13½ fl. oz.
Extract of orange flower	27 fl. oz.
Extract of rose	1 pint.
Extract of tuberose	3 pints.
Lily of the valley, syn	2 oz.

EAU DE LISBONNE.

Oil of lemon	2¼ oz.
Oil of orange	4½ oz.
Oil of rose	¼ oz.
Alcohol	5 qts.

MAGNOLIA (EXTRAIT DE MAGNOLIA).

Extract of orange flower	2 qts.
Extract of rose	4 qts.
Extract of tuberose	1 qt.
Extract of violet	1 qt.
Heliotropin	1 oz.
Oil of lemon	15 grains.
Oil of citronella, Java	1 dram.

LILY OF THE VALLEY.

Extract of jasmine	28 oz.
Extract of orange flower	7 oz.
Extract of rose	14 oz.
Extract of tuberose	7 oz.
Alcohol	28 oz.
Lily, syn. (Bush)	2 oz.

LILY OF THE VALLEY EXTRACT.

Extract of jasmine	3½ oz.
Extract of ylang-ylang (see below, p. 206)	½ oz.
Cardamon seed, crushed	75 grains.
Oil of orris	10 drops.
Lily of the valley, syn	1 dram.

Macerate for a week, and filter.

The amount of cardamon seed is to be weighed exactly;

should its odor still be too pronounced, extract of jasmine should be gradually added until the right aroma is obtained.

Bouquet a la Maréchale.

Tincture of ambergris	½ pint.
Tincture of musk	½ pint.
Extract of neroli	1 pint.
Extract of orange flower	1 qt.
Tincture of tonka	1 pint.
Tincture of vanilla	1 pint.
Tincture of orris root	1 pint.
Essence of vetiver	1 pint.
Essence of rose (triple)	1 qt.
Oil of clove	75 grains.
Oil of santal	75 grains.

A la Mode.

Extract of cassie	1 qt.
Extract of jasmine	1 qt.
Extract of orange flower	1 qt.
Extract of tuberose	1 qt.
Tincture of civet	1 pint.
Oil of bitter almond	75 grains.
Oil of nutmeg	60 grains.

A. Eau de Mille Fleurs.

Extract of cassie	1 pint.
Essence of rose	1 pint.
Extract of jasmine	1 pint.
Tincture of musk	6 fl. oz.
Extract of neroli	1 pint.
Extract of patchouly	1 pint.
Tincture of vanilla	2 oz.
Extract of violet	1 pint.
Essence of vetiver	2 oz.
Tincture of civet	6 fl. oz.
Citral	1 dram.
Oil of geranium	¾ oz.
Oil of lavender	¾ oz.
Musk xylene, 100%	1 dram.

B. Eau de Mille Fleurs.

Extract of cassie	1 pint.
Tincture of ambergris	½ pint.
Essence of cedar	½ pint.
Extract of jasmine	1 pint.
Tincture of musk	½ pint.
Extract of orange flower	1 pint.
Extract of rose	1 pint.
Extract of tuberose	1 pint.
Tincture of vanilla	½ pint.
Extract of violet	1 pint.
Essence of rose (simple)	1 qt.
Oil of bergamot	1¼ oz.
Oil of bitter almond	24 grains.
Oil of clove	24 grains.
Oil of neroli	24 grains.

C. Eau de Mille Fleurs a Palmarose.

Extract of cassie	6 fl. oz.
Essence of cedar	3 fl. oz.
Tincture of musk	3 fl. oz.
Extract of violet	6 fl. oz.
Oil of bergamot	1½ oz.
Oil of cedar	1¾ oz.
Oil of lemon	¼ oz.
Oil of lavender	¼ oz.
Oil of clove	¼ oz.
Oil of palmarosa	½ oz.
Alcohol	9 pints.

Fleurs de Montpellier.

Tincture of ambergris	10 fl. oz.
Tincture of musk	10 fl. oz.
Extract of rose	3 pints.
Extract of tuberose	3 pints.
Essence of rose (triple)	3 pints.
Oil of bergamot	1¾ oz.
Oil of clove	¼ oz.

FLEURS DES CHAMPS.

Extract of cassie......................	3½ oz.
Extract of jasmine....................	3½ oz.
Tincture of musk	3½ oz.
Tincture of tonka.....................	3 pints.
Tincture of orris root.................	7 oz.
Oil of geranium......................	1½ oz.
Oil of neroli.........................	1½ oz.
Oil of rose	⅞ oz.
Alcohol..............................	3 qts.

HUILE DE MILLE FLEURS.

(For perfuming hair oils and pomades.)

Oil of cinnamon......................	10 drops.
Oil of neroli.........................	20 drops.
Oil of rose	20 drops.
Oil of clove	—
Oil of orange	15 grains.
Oil of calamus.......................	20 drops.
Oil of geranium......................	150 grains.
Oil of lemon	½ oz.
Oil of bergamot......................	2½ oz.
Oil of verbena.......................	75 grains.

MUSK (EXTRAIT DE MUSC).

Musk xylene	1 dram.
Tincture of musk.....................	16 oz.
Tincture of civet.....................	8 oz.
Alcohol	4 qts.

MOUSSELINE.

Extract of cassie.....................	1 qt.
Extract of jasmine...................	1 qt.
Extract of rose......................	1 qt.
Extract of tuberose..................	1 qt.
Bouquet à la maréchale...............	2 qts.
Oil of santal........................	¾ oz.

MYRTLE (EXTRAIT DE MYRTHE).

Extract of jasmine	½ pint.
Extract of orange flower	1 qt.
Extract of rose	1 qt.
Extract of tuberose	1 qt.
Tincture of vanilla	1 qt.
Oil of myrtle	4 drams.

NARCISSUS (EXTRAIT DE NARCISSE).

Extract of jonquille	2 qts.
Extract of tuberose	3 qts.
Tincture of storax	1 oz.
Oil of narcisse, syn	3 oz.

NAVY'S NOSEGAY.

Extract of rose	1 qt.
Extract of orange flower	1 qt.
Essence of patchouly	3 fl. oz.
Extract of verbena	6 fl. oz.
Essence of vetiver	6 fl. oz.
Oil of bitter almond	150 grains.
Oil of citronella	1 dram.
Oil of nutmeg	75 grains.

NEW-MOWN HAY.

Tonka beans, in pieces	75 grains.
Orris root	150 grains.
Vanillin	8 grains.
Oil of bergamot	30 drops.
Oil of neroli	2 drops.
Oil of rose	2 drops.
Oil of lavender	2 drops.
Oil of clove	1 drop.
Patchouly herb	3 grains.
Benzoic acid	8 grains.
Alcohol	7½ oz.

Digest for two weeks, and filter.

PINK (EXTRAIT D'ŒILLET).

Extract of jasmine	2½ pints.
Extract of orange flower	2½ pints.
Extract of rose	5 pints.
Tincture of vanilla	20 fl. oz.
Oeillet (Naef)	2 oz.

ESSENCE OF SWEET PEA.

Extract of tuberose	1 qt.
Extract of orange flower	1 qt.
Extract of rose	1 qt.
Oil of sweet pea, syn. (Chiris)	3 oz.

POLYANTHUS.

Extract of rose	1 qt.
Extract of jasmine	1 pint.
Extract of violet	½ pint.
Tincture of musk	2½ fl. drachms.
Oil of neroli	¾ oz.
Oil of lemon	¾ oz.
Alcohol	2 qts.

EAU DU PORTUGAL.

Oil of bergamot	1 oz.
Oil of lemon	2¼ oz.
Oil of orange	½ lb.
Oil of rose	¼ oz.
Alcohol	5 qts.

QUEEN VICTORIA'S PERFUME.

Extract of cassie	10 fl. oz.
Extract of rose	5 pints.
Extract of orange flower	20 fl. oz.
Extract of tuberose	2½ pints.
Extract of violet	5 pints.
Tincture of civet	3 fl. oz.
Oil of bergamot	¾ oz.
Paquerette	4 oz.

PATCHOULY (EXTRAIT DE PATCHOULI).

Extract orange flower........................ 1 qt.
Oil of patchouly............................ 1½ oz.
Oil of rose................................150 grains.
Alcohol.................................... 1 gal.

ESSENCE OF RESEDA.

(Artificial, almost indistinguishable from the genuine.)

Tonka beans, in pieces....................... 30 grains.
Storax, liquid............................... 15 grains.
Orris root.................................. 1¾ oz.
Oil of neroli............................... 10 drops.
Oil of rose................................. 10 drops.
Oil of bitter almond........................ 2 drops.
Oil of bergamot............................. 20 drops.
Ambergris.................................. 15 grains.
Musk....................................... 8 grains.
Nettle herb................................ 30 grains.
Alcohol.................................... ½ lb.

Macerate for from one to two weeks, and filter.

RONDELETIA ODORATISSIMA.

Tincture of ambergris....................... 4¼ oz.
Tincture of musk........................... 4¼ oz.
Tincture of vanilla......................... 4¼ oz.
Oil of bergamot............................ 1 oz.
Oil of lavender............................ 2¼ oz.
Oil of clove............................... 1¼ oz.
Oil of rose................................ 75 grains.
Oil of sandal .... 2½ oz.
Alcohol.................................... 4 qts.

The odor of Rondeletia has not thus far been isolated, at least in Europe (the plant is indigenous to the Antilles). The oils of lavender and sandal together constitute the odor known in perfumery as Rondeletia. By increasing the quantity of the two oils, the strength of the perfume may be heightened.

ROYAL NOSEGAY.

Tincture of ambergris..................	2½ oz.
Extract of jasmine	1 qt.
Tincture of musk.....................	3 fl. oz.
Extract of rose	1 qt.
Tincture of vanilla	½ pint.
Extract of violet.....................	1 qt.
Essence of vetiver....................	½ pint.
Oil of bergamot......................	75 grains.
Oil of clove..........................	1¾ oz.

ROSE ODORS.

The art of perfumery has endeavored to fix this most magnificent of all odors, and we must confess that in this case it has succeeded in solving the problem in a manner unequalled in any other perfume. We are able to imitate not only the pure rose odor, but also those of its several varieties such as the tea rose, moss rose, etc., both as to character and intensity.

ROSA CENTIFOLIA, A (FINEST QUALITY).

Essence of rose (triple).....................	1 qt.
Rose pomade.............................	8 lb.
Alcohol.................................	5 qts.

ROSE, B (LESS FINE).

Oil of rose...............................	3½ oz.
Alcohol.................................	5 qts.

CHINA ROSE (ROSES JAUNES).

Essence of rose (triple)...............	2 qts.
Tincture of tonka.....................	8 oz.
Extract of jasmine....................	2 qts.
Extract of verbena....................	1 dram.
Oil of rose, syn. (Nadel)...............	4 oz.

Dog Rose (Eglantine).

Extract of cassie.........................	2½ pints.
Extract of jasmine.......................	2½ pints.
Extract of rose...........................	5 pints.
Essence of rose (triple)..................	2½ pints.
Rose Japon	4 oz.
Oil of neroli.............................	¼ oz.

Moss Rose (Rose Mousseuse).

Extract of rose...........................	2 qts.
Extract of orange flower.................	1 qt.
Essence of rose (triple)..................	1 qt.
Tincture of oak moss.....................	4 oz.
Tincture of musk.........................	½ lb.

Tea Rose (Rose Théa).

Extract of rose...........................	1 qt.
Extract of geranium......................	1 qt.
Extract of orange flower.................	½ pint.
Essence of rose (triple)..................	1 qt.
Extract of santal.........................	½ pint.
Tincture of orris root....................	½ pint.

White Rose (Roses Blanches).

Extract of rose...........................	2 qts.
Extract of jasmine.......................	2 qts.
Extract of cassie.........................	1 qt.
Oil of patchouly.........................	1 dram.
Otto of rose.............................	1 oz.

Rose Damascus.

Floressence rose	4 drams.
Floressence jasmine	4 drams.
Floressence cassie	1 dram.
Rose Damascus (Naef)....................	2 drams.
Otto of rose.............................	6 drams.
Cinthial	2 drams.
Musk xylene, 100%.......................	1 dram.
Alcohol	4 qts.

Spring Nosegay.

Extract of cassie........................	1 qt.
Viozone	4 oz.
Tincture of orris........................	1 pint.
Extract of jasmine.......................	1 qt.
Extract of orange flower..................	1 qt.
Floressence Mimosa	1 oz.
Musk xylene, 100%......................	1 dram.

Suave.

Extract of cassie........................	1 qt.
Tincture of civet........................	3 oz.
Extract of jasmine.......................	1 qt.
Tincture of musk........................	¼ pint.
Extract of rose..........................	1 qt.
Extract of tuberose......................	1 qt.
Coumarin (Monsanto)	4 drams.
Oil of bergamot.........................	½ oz.
Oil of clove.............................	30 grains.
Oil of mace.............................	30 grains.

Heliotrope Bouquet (Fleurs Solsticiales).

Extract of cassie........................	13½ fl. oz.
Tincture of ambergris....................	5 fl. oz.
Extract of jasmine.......................	2½ pints.
Tincture of musk........................	5 fl. oz.
Extract of rose..........................	5 pints.
Extract of violet........................	2½ pints.
Extract of verbena.......................	13½ fl. oz.
Essence of rose (triple)..................	2½ pints.
Heliotropine	4 oz.
Musk xylene, 100%......................	1 dram.

Bouquet de Stamboul.

Extract of rose..........................	2½ pints.
Extract of cassie........................	1 qt.
Extract of jasmine.......................	1 qt.
Extract of tuberose......................	1 pint.
Tincture of civet........................	½ pint.
Opopanol	4 oz.

Syringa.

Extract of jasmine..........................	2 qts.
Extract of orange flowers...................	2 qts.
Fleur d'Orange	2 oz.
Syringat	2 oz.
Cyclosia	1 dram.
Musk xylene, 100%.......................	1 dram.

Tulipe Odoriférante.

Extract of orange flowers...................	6 fl. oz.
Extract of jasmine..........................	1 qt.
Extract of rose............................	1 pint.
Extract af tuberose.......................	1 qt.
Tincture of orris root......................	1 qt.
Aurantiol (Givaudan)......................	1 oz.
Oil of neroli..............................	30 grains.

Hungarian Water (Eau Hongroise).

Extract of orange flower....................	1 pint.
Essence of rose (triple)....................	1 pint.
Oil of lemon..............................	1 oz.
Oil of melissa.............................	1 oz.
Oil of peppermint.........................	30 grains.
Oil of rosemary...........................	2 oz.
Alcohol	5 qts.

Bouquet de Virginie.

Tincture of musk..........................	4 oz.
Extract of orange flower...................	1 qt.
Extract of santal..........................	1 pint
Tincture of tonka.........................	1 qt.
Tincture of vanilla........................	4 oz.
Essence of rose (triple)....................	1 pint.
Tilleul (Naef)	2 oz.
Oil gardenia, simile.......................	1 oz.
Fougere (V. C.)...........................	1 oz.
Tincture of orris..........................	1 pint.
Musk ambrette	1 dram.

VIOLET (VIOLETTE).

Floressence violet	1 oz.
Floressence cassie	1 dram.
Tincture of orris root...................	1 qt.
Irisolette	4 drams.
Heliotropin	1 dram.
Alcohol	5 qts.

VERBENA, A (EXTRAIT DE VERVEINE).

Citral	½ oz.
Oil of lemon..........................	1 oz.
Oil of bergamot.......................	½ oz.
Alcohol	4 qts.

A cheap and pleasant perfume. The following is far superior.

VERBENA, B.

Oil of lemon..........................	4 oz.
Oil of verbena........................	4 oz.
Oil of bergamot.......................	2 oz.
Extract of orange flower...............	2 lb.
Extract of rose.......................	3 lb.
Extract of tuberose...................	2 lb.
Alcohol	5 qts.

This "Extract of Verbena, B," is a modification of that given previously, on page 172.

EXTRAIT DE VERVEINE, C.

Extract of orange flower...............	1 qt.
Extract of rose.......................	1 qt.
Extract of tuberose...................	1 qt.
Oil of lemon..........................	1 oz.
Oil of verbena........................	¾ oz.
Oil of bergamot.......................	1 oz.
Alcohol	1 qt.

MOUNTAIN VIOLETS.

Extract of cassie.....................	13½ fl. oz.
Extract of jasmine....................	13½ fl. oz.
Extract of rose.......................	13½ fl. oz.
Extract of violet.....................	2 qts.
Tincture of orris root.................	13½ fl. oz.
Ionone, 100%......................	2 drams.

VOLCAMERIA.

Extract of jasmine....................	1 pint.
Extract of rose.......................	1 qt.
Extract of tuberose...................	2 qts.
Extract of violet.....................	2 qts.
Tincture of musk....................	½ pint.

FOREST BREEZE (PINE-NEEDLE ODOR.)

Oil of turpentine....................	14 oz.
Oil of lavender......................	1½ oz.
Citral	¾ oz.
Alcohol............................	5 qts.

The oil of turpentine must be clear like water, and most carefully rectified. If it can be obtained of good quality, the oil distilled from the leaves or needles of Pinus sylvestris, commonly known as pine-needle oil or fir-wool oil, is to be preferred for this purpose. Still better is the oil obtained from Pinus Pumilio.

WEST END.

Extract of rose......................	1 qt.
Tincture of ambergris.................	½ pint.
Extract of jasmine....................	1 qt.
Tincture of musk....................	½ pint.
Extract of tuberose...................	1 qt.
Extract of violet.....................	1 qt.
Essence of rose (triple)...............	3 pints.
Oil of bergamot......................	1 oz.
Oil of lemon........................	75 grains.
Oil of lavender......................	75 grains.

FLEUR D'AMOUR.

Extract of cassie........................ 1 qt.
Tincture of ambergris.................... 4 oz.
Extract of lavender...................... 1 pint.
Extract of orange flower................. 1 qt.
Extract of rose.......................... 1 qt.
Orchidee (Naef) .... 4 drams.
Essence of vetiver....................... 1 pint.

FLOWERS OF THE ISLE OF WIGHT.

Extract of rose.......................... 1 qt.
Extract of santal........................ 2 qts.
Tincture of orris root................... 1 qt.
Essence of vetiver....................... 1 pint.

YACHT CLUB.

Extract of cassie........................ 6 fl. oz.
Extract of jasmine....................... 1 qt.
Extract of orange flower................. 2 qts.
Extract of santal........................ 2 qts.
Wistaria (V. C.)......................... 2 oz.
Essence of rose (triple)................. 1 qt.
Benzoic acid, sublimed................... 1½ oz.

The characteristic odor of this perfume depends upon the volatile oil adhering to the sublimed benzoic acid; for this reason no other benzoic acid should be used than that obtained by sublimation.

YLANG YLANG.

Extract of jasmine....................... 3 qts.
Essence of rose (triple)................. 1 qt.
Heliotropine .... 1 oz.
Tincture of styrax....................... 1 oz.
Oil of neroli............................ 75 grains.
Oil of Ylang Ylang....................... 2 oz.

APPENDIX.

The great majority of the above-described perfumes are made with extracts prepared from pomades; hence their cost of production is considerable and the selling-price high. For the requirements of the middle classes, quite fragrant perfumes are manufactured by dissolving the cheaper essential oils in ordinary alcohol, and various new odors can be obtained by mixing several of them. The extracts made with cheap oils are well suited to this purpose. The oils most frequently used for such articles are those of bergamot, lemon, orange peel, lavender flowers (French), lemon grass, nutmeg, clove, and santal. The alcohol must be free from fusel oil and have a strength of at least 70% Tralles.

Oils with not very intense odor are generally used in the proportion of about 2 to 2½ ounces to the quart of alcohol; half that quantity will suffice for strong-scented oils such as those of lemon-grass, clove, and nutmeg.

From these simple solutions an experienced manufacturer can produce very nice perfumes by mixing them in due proportions; they are comparatively cheap, and sometimes they yield relatively more profit than the finest articles, whose contents and containers generally represent a considerable outlay on the part of the manufacturer.

Modern practice, however, tends more to the use of synthetic products even for the higher class of perfumes. The following are examples of this tendency:

HELIOTROPE.

Oil of ylang-ylang........................	20 drops.
Geraniol.................................	10 drops.
Benzaldehyde............................	2 drops.
Heliotropin.............................	35 grains.
Vanillin.................................	6 grains.
Coumarin................................	4 grains.
Tincture of musk (xylene 100%)...........	40 grains.
Cologne spirit (95%) enough to make......	1 qt.

LILAC.

Aubepine, liquid	1 dram.
Cyclosia	1 dram.
Oil of bergamot..........................	1 dram.
Oil of jasmine, syn.......................	1 dram.
Oil of jacinthe, syn. Verley..............	2 drams.
Terpineol	1 oz.
Vanillin	12 grains.
Cologne spirit (90%), enough to make.......	1 gal.

MIGNONETTE.

Floressence mignonette	1 oz.
Oil of neroli.............................	½ dram.
Oil of jasmine (synthetic).................	½ dram.
Ionone, 100%	1 dram.
Oil of bitter orange......................	15 drops.
Cologne spirits (90%), enough to make......	4 qts.

WOOD VIOLET.

Solution of ionone (1 in 30, in 60% alcohol)......	2 pints.
Solution of oil of orris (concrete) (1 in 60, in 60% alcohol)................................	2 pints.
Oil of bergamot...........................	1 dram.
Solution of artificial musk (1%, in 95% alcohol)..	2 oz.

MUGUET.

Oil of jasmine (synthetic).................	1 dram.
Oil of ylang-ylang (synthetic).............	3 drams.
Solution of heliotropin (1%, in 60% alcohol)	8 oz.
Solution of terpineol (1%, in 60% alcohol)......	20 oz.

SWEET HAWTHORN.

Anisic aldehyde...........................	1 dram.
Oil of linaloe............................	3 drams.
Oil of jasmine (synthetic).................	1 dram.
Oil of neroli.............................	1 dram.
Solution of artificial musk (1%, in 95% alcohol).	20 oz.
Alcohol (90%) enough to make................	160 oz.

LILY OF THE VALLEY.

Oil of bois de rose...........................	90 minims.
Oil of ylang-ylang...........................	30 minims.
Oil of jasmine (synthetic)...................	15 minims.
Lily of the valley, syn. (de Laire).........	90 minims.
Tincture of musk............................	30 drops.
Cologne spirit (90%) enough to make..........	1 qt.

VERBENA.

Oil of neroli...............................	1 oz.
Oil of verbena..............................	5 oz.
Oil of lemon................................	6 oz.
Oil of rose.................................	1 oz.
Oil of geranium.............................	2 oz.
Cologne spirit (90%) enough to make..........	3 qts.

TREFLE INCARNAT.

Oil of bergamot.............................	6 fl. oz.
Oil of rose, Bulgarian......................	6 fl. drams.
Oil of hyacinth.............................	3 fl. drams.
Oil of neroli...............................	40 minims.
Oil of ylang-ylang..........................	1 fl. dram.
Oil of white thyme..........................	40 minims.
Oil of vetivert.............................	1 fl. dram.
Amyl salicylate.............................	2½ fl. oz.
Artificial musk (xylene 100%)...............	10 drams.
Vanillin....................................	1 oz.
Tincture of civet...........................	1 pint.
Rose water..................................	2 qts.
Cologne spirit..............................	18 qts.

MAGNOLIA.

Extract of tuberose.........................	12 fl. oz.
Extract of jonquil..........................	1½ fl. oz.
Extract of orange flower....................	1½ fl. oz.
Oil of rose (synthetic).....................	15 minims.
Oil of jasmine (synthetic)..................	8 minims.
Oil of neroli (synthetic)...................	8 minims.
Methyl anthranilate.........................	3 minims.

CORYLOPSIS.

Oil of rose (synthetic).........................	1 fl. oz.
Oil of ylang-ylang............................	1½ fl. oz.
Oil of patchouly.............................	1 fl. dram.
Tincture of benzoin.........................	2½ fl. oz.
Geraniol....................................	1 fl. dram.
Muguet (synthetic), Muhlenthaler........	1 fl. dram.
Cinnamic alcohol...........................	1 fl. dram.
Extract of jasmine..........................	1 qt.
Cologne spirit..............................	1 gal.

NEW-MOWN HAY.

Coumarin	45 grains.
Vanillin....................................	30 grains.
Heliotropin................................	15 grains.
Solution of Ionone, 10%....................	15 minims.
Oil of rose.................................	8 minims.
Oil of neroli...............................	8 minims.
Oil of patchouly............................	2 minims.
Terpineol..................................	8 minims.
Tincture of benzoin.........................	6 fl. dram.
Essence of tuberose.........................	3 fl. oz.
Essence of jasmine..........................	6 fl. oz.
Alcohol (80%) enough to make...............	1 qt.

CHAPTER XVII.

AMMONIACAL AND ACID PERFUMES.

A. AMMONIACAL PERFUMES.

AMMONIA (ammonia water) has a disagreeable odor and exerts a very caustic effect on the lachrymal glands. Despite these properties, ammonia, in a highly dilute condition and mixed with other aromatics, finds manifold application in perfumery and serves particularly for the manufacture of the so-called smelling salts, or inexhaustible salts, used for filling smelling bottles.

The liquid or caustic ammonia, however, is not so suitable for the purposes of the perfumer as the carbonate of ammonia, which when pure forms colorless crystals usually covered with a white dust (consisting of bicarbonate of ammonia); these, undergoing gradual decomposition, give off the odor of ammonia and hence are more lasting in smelling bottles than the pure liquid ammonia.

The main essential for both of these substances is purity. Caustic ammonia as well as carbonate of ammonia are now obtained on a large scale from "gas liquor," but the crude products always retain some of the penetrating odor of coal tar which renders them valueless for the purposes of the perfumer. We must, therefore, make it a rule to use nothing but perfectly pure materials which, moreover, are easily to be had in the market.

INEXHAUSTIBLE SALT (SEL INÉPUISABLE).

Oil of bergamot	24 grains.
Oil of lavender	45 grains.
Oil of mace	24 grains.
Oil of clove	24 grains.
Oil of rosemary	45 grains.
Water of ammonia	1 qt.

The aromatics are placed in a bottle, the ammonia is added, and the bottle vigorously shaken; the solution is soon effected, and the turbid liquid can be at once filled into bottles.

According to the material from which the containers are made, different methods must be adopted. It is necessary to give the liquid such form as to prevent its flowing out when the vessel is inverted; this is important, as the bottles are often carried in dress pockets and the ammonia destroys most colors. As a rule the vessels are filled with indifferent porous substances which are moistened with the perfume. If the container is made of box wood, ivory, porcelain, or some other opaque material, it is filled with fibres of asbestos or with very small pieces of sponge, and as much perfume is poured in as the substance can take up; the vessels are then inverted into a porcelain plate and allowed to drain, and are finally closed with a loose plug of cotton. If the container is transparent, it is better to use, instead of the asbestos or sponge which do not look neat, either small pieces of white pumice stone, powdered glass, small white glass beads, or crystals of sulphate of potassium which is insoluble in the perfume.

WHITE SMELLING SALT (SEL BLANC PARFUMÉ).

While the first-named ammoniacal preparation is called a salt, it is really nothing but perfumed caustic ammonia; but white smelling salt is what its name indicates and can be perfumed as desired by the consumer; but as only certain scents harmonize with ammonia, not every odor can be employed; the most appropriate are oils whose odor resembles that of rose, and the oils of nutmeg and cinnamon.

Mix in a large porcelain jar—

Carbonate of ammonia.................. 2 lb.
Caustic ammonia...................... 1 lb.

Cover the jar and leave it at rest. After some days the mixture will have changed into a firm mass of monocarbonate of ammonia which is rubbed to a coarse powder, perfumed, and filled into bottles. The above quantities require:

Oil of bergamot........................ 15 grains.
Oil of lavender........................ 15 grains.
Oil of nutmeg......................... 8 grains.
Oil of clove.......................... 8 grains.
Oil of rose .... 8 grains.
Oil of cinnamon....................... 75 grains.

The oils are poured into a mortar and rubbed up with about one-tenth of the salt; of this perfumed salt enough is added to the several portions of the mass, and triturated until the odor is equally distributed. For cheaper smelling salts oils of geranium and cassia may be substituted for the oils of rose and cinnamon.

· PRESTON SALT (SEL VOLATIL).

In this perfume ammonia is continually generated; the salt is prepared by mixing chloride of ammonium or sal-ammoniac in fine powder with freshly slaked lime. Fine or cheap perfume is added, according to the grade desired. The mixture of sal-ammoniac and slaked lime continually develops small amounts of ammonia—it takes a long time until the decomposition is complete, and for this reason a bottle filled with Preston salt retains the odor of ammonia for several years.

EAU DE LUCE.

This is the only ammoniacal perfume used in a liquid form. It is made according to the following formula:

Tincture of ambergris.................... 10½ oz.
Tincture of benzoin.................... ½ lb.
Oil of lavender........................150 grains.
Water of ammonia.................... 1½ lb.

The tinctures are mixed with the ammonia by agitation and immediately filled into bottles; the liquid should have a milky appearance. At times 150 grains of white soap is added which aids in imparting to the liquid the desired milky appearance. In fine eau de Luce the odor of ambergris should predominate; this can be easily effected by increasing the amount of tincture of ambergris.

B. ACID PERFUMES.

As there is a group of perfumes which is distinguished by their characteristic odor of ammonia and which we have therefore called ammoniacal, so there is an important series of arti-

cles containing acetic acid which are used cosmetically as
so-called toilet vinegars, and in some washes.

Ordinary vinegar, *i.e.*, water containing four to six per
cent of acetic acid, has, as is well known, a not unpleasant
refreshing odor and a pure acid taste. Pure acetic acid, now
made in large quantities and of excellent quality, is known
commercially as glacial acetic acid. In commerce, it is customary to designate any acetic acid containing 85 or more
per cent of the absolute acid, as glacial acetic acid. In chemical or pharmacopœial nomenclature, however, the glacial acid
is meant to be as near 100% as possible. In perfumery, an 85%
acid is sufficiently strong. It forms a colorless liquid with a
narcotic odor and an intensely acid taste; it congeals into
glassy crystals at a temperature of 8.5° C. (47° F.). The latter
property is of importance as showing the purity of the acid.
Concentrated acetic acid, like alcohol, dissolves aromatic substances, with which it forms perfumes which differ from those
made with alcohol mainly by their peculiar refreshing after-
odor which is due to the acetic acid.

Acetic acid can be saturated with various odors and thus
furnish fine perfumes; but for so-called toilet vinegars which
are used as washes the acetic acid must be properly diluted,
since the concentrated acid has pronounced caustic properties,
reddens the skin, and may even produce destructive effects
on sensitive parts such as the lips.

AROMATIC VINEGAR (VINAIGRE AROMATIQUE).

Glacial acetic acid...................... 2 lb.
Camphor............................... 4¼ oz.
Oil of lavender........................ ¾ oz.
Oil of mace............................150 grains.
Oil of rosemary ....150 grains.

Instead of the perfumes here given, finer odors may be
employed for the production of superior toilet vinegars; thus

we find vinaigre ambré, au musc, à la violette, au jasmin, etc., according to the perfume used. As concentrated acetic acid dissolves most aromatic substances the same as alcohol, all alcoholic perfumes may have their counterparts in acetic acid; but the aromatics should never be added in so large amount as to mask the characteristic odor of the acetic acid. A very pleasant vinegar may be produced by combining an alcoholic with an acid perfume, as in the following:

SPICED VINEGAR (VINAIGRE AUX ÉPICES).

1. Macerate:
   Leaves of geranium, lavender, pepper-
   mint, rosemary, and sage, of each.... 1 oz.
   In alcohol of 80%.................. 1 lb.
2. Macerate:
   Angelica root, calamus root, camphor,
   mace, nutmeg, cloves, of each...... ½ oz.
   In glacial acetic acid.............. 2 lb.

for two weeks, mix the liquids, and filter them into a bottle which should not be completely filled. The longer this mixture is allowed to season in the bottle, the finer will be the aroma; for in the course of time the alcohol and acetic acid react on each other and form acetic ether, which likewise possesses a pleasant aromatic odor.

Certain aromatic vinegars, like ammoniacal perfumes, are filled into smelling bottles containing the same porous substances for their absorption, namely, sponge, pumice stone, crystals of potassium sulphate, etc.

FORMULAS FOR TOILET VINEGARS.

VINAIGRE A LA ROSE.

Essence of rose (triple)................. 10½ oz.
White-wine vinegar.................... 1 qt.

This should be colored a pale rose tint with one of the dye-stuffs to be enumerated hereafter. The use of true wine vinegar is to be recommended for this and all the following toilet vinegars, as the œnanthic ether it contains has a favorable effect on the fineness of the odor.

VINAIGRE AUX FLEURS D'ORANGES.

Extract of orange flower	7 oz.
White-wine vinegar	1 qt.

This is usually left colorless.

VINAIGRE AUX VIOLETTES.

Extract of cassie	8 oz.
Extract of orange flower	3½ oz.
Tincture of orris root	5½ oz.
Essence of rose (triple)	5½ oz.
White-wine vinegar	1 qt.

VINAIGRE DE QUATRE VOLEURS.

Leaves of lavender, peppermint, rue, rosemary, and cinnamon, of each	3¼ oz.
Calamus, mace, nutmeg, of each	150 grains.
Camphor	¾ oz.
Macerated in alcohol	7 oz.
And acetic acid	4¾ lb.

PREVENTIVE VINEGAR (VINAIGRE HYGIÉNIQUE).

Benzoin	2¼ oz.
Lavender	¾ oz.
Cloves	150 grains.
Marjoram	¾ oz.
Cinnamon	150 grains.
Alcohol	1 qt.
White-wine vinegar	2 qts.

Macerate the solids with the alcohol and vinegar.

VINAIGRE DE COLOGNE.

Cologne water	1 qt.
Glacial acetic acid	1¾ oz

As this vinegar is made by mixing an alcoholic perfume with acetic acid, so all other alcoholic perfumes may be employed for a like purpose; but the quantities must be determined by experiment, for the various aromatics differ in the intensity of their odor.

VINAIGRE ÉTHERÉ.

Glacial acetic acid....................	14 oz.
Acetic ether.........................	1½ oz.
Nitrous ether	¾ oz.
Water...............................	5 qts.

The water should be added after the ethers have been dissolved in the glacial acetic acid.

VINAIGRE DE LAVANDE.

Lavender water......................	4 qts.
Rose water..........................	1 pint.
Glacial acetic acid....................	½ lb.

To be stained a bluish color with indigo-carmine.

ORANGE-FLOWER VINEGAR.

Orange-flower water..................	4 qts.
Glacial acetic acid...................	7 oz.

MALLARD'S TOILET VINEGAR.

Tincture of benzoin...................	1½ oz.
Tincture of tolu......................	1½ oz.
Oil of bergamot......................	150 grains.
Oil of lemon.........................	150 grains.
Oil of neroli........................	30 grains.
Oil of orange peel....................	½ oz.
Oil of lavender......................	15 grains.
Oil of rosemary......................	15 grains.
Tincture of musk.....................	15 grains.
Concentrated acetic acid	21 oz.
Alcohol.............................	4¾ lb.

TOILET VINEGAR (FRENCH FORMULA).

Oil of bergamot.......................	30 grains.
Oil of lemon........................	30 grains.
Oil of rose...........................	8 drops.
Oil of neroli.........................	5 drops.
Benzoin..............................	75 grains.
Vanillin.............................	15 grains.
Concentrated acetic acid...............	½ oz.
Alcohol.............................	½ lb.

Macerate for two weeks, and filter.

VINAIGRE POLYANTHE.

Glacial acetic acid....................	7 oz.
Tincture of benzoin...................	1¾ oz.
Tincture of tolu......................	1¾ oz.
Oil of neroli.........................150 grains.	
Oil of geranium......................150 grains.	
Water	2 qts.

To be stained with tincture of krameria (rhatany).

CHAPTER XVIII.

DRY PERFUMES.

As a matter of course, dry perfumes are of greater antiquity than fluid; aromatic substances require merely to be dried in order to retain their fragrance permanently. The oldest civilized people known in history—Egyptians, Assyrians, Persians, Babylonians, and the Jews, as numerous passages in the Bible prove—used dried portions of plants, leaves, flowers, and resins as perfumes and incense.

To this day there is kept up quite a trade in Valeriana celtica, a strong-scented Alpine plant, and in powdered amber, with the Orient, where they are used for scent bags and incense respectively.   The Catholic Church retains to the present time the Jewish rite of burning incense, and in our museums will be found urns, taken from Egyptian graves, from which pleasant odors escape even now after nearly four thousand years, owing to the aromatic resins with which they are filled.   It is said, too, that the delightful volatile odors of our handkerchief perfumes were first prepared by an Italian named Frangipanni conceiving the idea of treating a dry mixture of different aromatic plants with alcohol and thus imparting the odor they contained to the latter.

Not all aromatics can be made into sachet powders; it is well known that the delightful odor of violets changes into a positively disagreeable smell when the flowers are dried, and the same remark applies to the blossoms of the lily of the valley, mignonette, lily, and most of our fragrant plants.   On the other hand, some portions of plants, especially those in which the odorous principle is contained not only in the flower but in all parts of the plant, as in the mints, sage, and most Labiatæ, remain fragrant for a long time after drying and hence can be employed for sachets.   Besides the plants named, lavender, rose leaves, the leaves of the lemon and orange tree, Acacia farnesiana, patchouly herb, and some other plants continue fragrant after drying.

Any vegetable substance to be used for sachets must be completely dried so as to prevent mould.   The drying should be effected in a warm, shady place, sometimes in heated chambers; direct sunlight and excessive heat injure the strength of the odor, a portion of the aromatics becoming resinified and volatilized.   If artificial heat is employed, a temperature between 40 and 45° C. (104–113° F.) is most suitable.

The external form of this class of preparations varies of course with the public for which it is intended.   Expensive sachets are sold in silk bags with different ornamentation; those intended for the Orient are generally put up as small silk cushions richly ornamented with gold and colors to suit Oriental taste. Cheap sachets are sold in envelopes or in round boxes. It is customary to have the ingredients ground or finely powdered, for which purpose small hand-mills will generally suffice.

CHAPTER XIX.

FORMULAS FOR DRY PERFUMES (SACHETS).

CEYLON SACHET POWDER.

Mace...............................	23 oz.
Patchouly...........................	28 oz.
Vetiver root........................	35 oz.
Rose leaves.........................	35 oz.

CYPRIAN SACHET POWDER.

Cedar wood.........................	2 lb.
Rhodium............................	2 lb.
Santal wood........................	2 lb.
Oil of rhodium.....................	½ oz.

The oil is mixed with the finely powdered or rasped woods and distributed in the mass by trituration.

FIELD FLOWER SACHET POWDER.

Calamus root	1 lb.
Caraway	½ lb.
Lavender	1 lb.
Marjoram...........................	½ lb.
Musk	30 grains.
Cloves.............................	2¾ oz.
Peppermint.........................	½ lb.
Rose leaves........................	1 lb.
Rosemary...........................	3½ oz.
Thyme	½ lb.

FRANGIPANNI SACHET POWDER.

Musk .... 1 oz.
Sage .... ½ lb.
Santal wood......................... ½ lb.
Orris root.......................... 6 lb.
Vetiver .... ½ lb.
Civet .... ¼ oz.
Oil of neroli .... 75 grains.
Oil of santal .... 75 grains.
Oil of rhodium...................... 75 grains.

HELIOTROPE SACHET POWDER.

Musk............................... ¼ oz.
Rose leaves......................... 2 lb.
Tonka beans......................... 1 lb.
Vanilla............................ ½ lb.
Orris root.......................... 4 lb.
Heliotropin........................ 8 oz.

INDIAN SACHET POWDER.

Santal wood......................... 3½ oz.
Orris root.......................... 21 oz.
Cinnamon........................... 10½ oz.
Oil of lavender..................... 75 grains.
Cloves............................. 30 grains.
Oil of rose........................150 grains.

LAVENDER SACHET POWDER.

Benzoin .... 1 lb.
Lavender flowers.................... 4 lb.
Oil of lavender..................... 1 oz.
Oil of rose......................... 75 grains.

MARSHAL SACHET POWDER.

Cassia .... ½ lb.
Musk............................... 75 grains.
Cloves............................. ½ lb.
Rose leaves........................ ½ lb.
Santal wood........................ 1 lb.
Orris root .... 1 lb.

MILLE FLEURS SACHET POWDER.

Benzoin.................................... 1 lb.
Lavender................................. 1 lb.
Musk..................................... 30 grains.
Cloves................................... 4¼ oz.
Allspice................................. 2½ oz.
Rose leaves.............................. 1 lb.
Santal wood............................. 4¼ oz.
Tonka beans............................. 4¼ oz.
Vanilla.................................. 4½ oz.
Orris root............................... 1 lb.
Civet.................................... 30 grains.
Cinnamon................................ ½ oz.

MUSLIN SACHET POWDER.

Benzoin.................................. ½ lb.
Santal wood............................. 1 lb.
Thyme .... 1 lb.
Orris root............................... 1 lb.
Vetiver root............................. 2 lb.
Oil of geranium......................... 75 grains.

OLLA PODRIDA.

This name is applied in Spain to a dish prepared from various remnants of food. The olla podrida of the perfumer is made from the remnants of the aromatic vegetable substances after their extraction with alcohol, petroleum ether, etc. Although vanilla, cinnamon, nutmeg, etc., be repeatedly extracted, they still retain their characteristic odor, though somewhat fainter, and thus they can be used with advantage for sachet powders intended for filling bags, cushions, etc. If mixed in corresponding proportions, they can be made use of for all the sachets here enumerated. No definite formula can be given for a peculiar dry perfume to be called Olla podrida; the olfactory organ is the best guide.

PATCHOULY POWDER.

Patchouly herb.......................	2 lb.
Oil of patchouly......................	30 grains.
Musk................................	15 grains.

The musk is rubbed up with gradually increased quantities of the patchouly herb and with the addition of the oil of patchouly; the intimate mixture of the powder saturated with musk and oil of patchouly and the rest of the powder is effected by prolonged stirring of the two powders in a large vessel. The same process is followed with all other dry powders in which a small amount of a solid with intense odor or of an essential oil is to be mixed with a large quantity of powder.

PERSIAN SACHET POWDER.

Musk................................	30 grains.
Rose leaves..........................	1 lb.
Tonka beans.........................	3½ oz.
Orris root	2 lb.
Oil of nutmeg........................	75 grains.
Oil of clove..........................	75 grains.
Oil of rose...........................	150 grains.
Oil of cinnamon......................	75 grains.

PORTUGAL POWDER.

Lemon peels.........................	1 lb.
Orange peels	2 lb.
Orris root............................	1 lb.
Cinnamon............................	3½ oz.
Oil of lemon grass....................	150 grains.
Oil of neroli.........................	150 grains.
Oil of orange	2½ oz.

POTPOURRI.

Many widely differing perfumes are sold in the market under this name; a good formula for its preparation is the following:

Lavender.	1 lb.
Cloves	2½ oz.
Allspice.	2½ oz.
Rose leaves.	1 lb.
Reseda.	1¾ oz.
Orris root	½ lb.
Vanilla.	150 grains.
Cinnamon.	1¾ oz.
Sand, or table salt, etc.	1 lb,

The admixture of fine white sand, table salt, or powdered glass or marble, etc., is made merely for the purpose of increasing the weight.

ROSE SACHET POWDER, A.

Geranium herb	3½ oz.
Rose leaves.	2 lb.
Santal wood.	1 lb.
Oil of rose.	½ oz.

ROSE SACHET POWDER, B.

Rose leaves.	2 lb.
Santal wood	1 lb.
Oil of rose.	1 oz.

SANTAL POWDER,

which is simply finely rasped santal wood, is also sometimes sold as rose sachet powder when it has received an addition of some oil of geranium.

VIOLET SACHET POWDER.

Benzoin.	½ lb.
Musk.	30 grains.
Orange flowers.	1¾ oz.
Rose leaves.	1 lb.
Orris root	2 lb.
Ionone 100%.	1 dram.
Heliotropin.	1 oz.

VIOLET SACHET POWDER.

Orris root, powdered.................	1 lb.
Musk.............................	8 grains.
Vanillin...........................	30 grains.
Oil of rose........................	25 drops.
Oil of petit grain..................	5 grains.
Ionone, 100%.....................	25 drops.

Mix intimately in a porcelain mortar.

VERBENA SACHET POWDER.

Lemon peels.......................	1 lb.
Caraway..........................	½ lb.
Orange peels......................	1 lb.
Oil of bergamot...................	1¾ oz.
Oil of lemon......................	1¾ oz.
Oil of verbena....................	75 grains.

VETIVER SACHET POWDER.

Vetiver root.......................	2 lb.
Musk.............................	15 grains.
Civet.............................	20 grains.

CHAPTER XX.

THE PERFUMES USED FOR FUMIGATION.

ACCORDING to the use made of them, perfumes for fumigation may be divided into two groups: those which develop their fragrance on being burned, and those which do so on being merely heated. The former group includes pastils and ribbons; the latter, fumigating powders and waters.

FUMIGATING PASTILS.

French—Pastilles fumigatoires; *German*—Räucherkerzen.

Pastils consist in the main of charcoal to which enough saltpetre is added to make the lighted mass glow continuously and leave a pure white ash. To this mass are added various aromatic substances which are gradually volatilized by the heat and fill the surrounding air with their perfume. It is important to observe that only ordinary saltpetre (nitrate of potassium) is to be used for this purpose, and not the so-called Chili saltpetre (nitrate of sodium) which becomes moist in the air. For ordinary pastils finely rasped fragrant woods such as cedar or santal are frequently employed. During the slow combustion, however, the wood gives off products of a pungent or disagreeable odor such as acetic acid and empyreumatic products, which lessen the fragrance. Fine pastils are composed of resins and essential oils and are usually formed into cones two-fifths to four-fifths of an inch high, by being pressed in metal moulds.

Fumigating pastils are manufactured as follows. Each solid ingredient is finely powdered by itself, and the necessary quantities are then put into a wide porcelain dish and intimately mixed with a flat spatula. In order to confine the dust, the dish is covered with a cloth during this operation. The mixture being completed, the essential oils are added, together with enough mucilage of acacia to form a plastic mass to be kneaded with the pestle, and which after drying will have a sufficiently firm consistence.

PASTILLES ORIENTALES.

Charcoal	1½ lb.
Saltpetre	3½ oz.
Benzoin	½ lb.
Powdered amber	3½ oz.
Tolu balsam	2¾ oz.

The charcoal for this and all other pastils should be made from soft woods (willow, poplar, etc.). The characteristic of these pastils is the amber they contain (the offal from manufactories is used) and which on ingition gives off a peculiar odor much prized in the Orient, rather than in Europe or America.

PASTILLES DU SÉRAIL.

Charcoal............	1½ lb.
Saltpetre.........	3½ oz.
Benzoin............................	½ lb.
Santal wood.......................... .	5½ oz.
Opium..............................	1¾ oz.
Tolu balsam	2¾ oz.

This formula is here given as usually quoted. It may be stated, however, that the opium may be omitted entirely, as it neither contributes to the fragrance nor produces, by being burned in this manner, any of the supposed exhilarating or intoxicating effects which it may produce when used in other forms or employed in other ways.

BAGUETTES ENCENSOIRES (FUMIGATING PENCILS).

Benzoin.....	14 oz.
Charcoal.............................	1¾ oz.
Peru balsam..........................	1 oz.
Storax..............................	2 oz.
Shellac............................	3½ oz.
Olibanum...........................	5½ oz.
Civet	75 grains.
Oil of bergamot......................	1 oz.
Oil of orange peel.	1 oz.
Oil of santal...............	¾ oz.

Melt the benzoin, charcoal, shellac, and olibanum in a bright iron pan at the lowest possible heat; take the pan from the fire and add the other ingredients, heat being again applied from time to time to keep the mass in a liquid state. The plastic mass is rolled out on a marble slab into rods the thickness of a lead pencil. Such a pencil need be but lightly passed over a hot surface to volatilize the aromatics it contains.

PASTILLES ODORIFÉRANTES.

Charcoal	2 lb.
Saltpetre	3½ oz.
Benzoin	1½ lb.
Cloves	7 oz.
Tolu balsam	7 oz.
Vanilla	7 oz.
Vetiver root	7 oz.
Cinnamon	3½ oz.
Oil of neroli	150 grains.
Oil of santal	¾ oz.

This and the following formula give the finest mixtures for pastils.

PASTILLES ENBAUMÉES.

Charcoal	2 lb.
Saltpetre	2¾ oz.
Benzoic acid, sublimed	1 lb.
Musk	15 grains.
Civet	15 grains.
Oil of lemon grass	30 grains.
Oil of lavender	15 grains.
Oil of clove	15 grains.
Oil of rose	15 grains.
Oil of thyme	30 grains.
Oil of cinnamon	30 grains.

POUDRE D'ENCENS (INCENSE POWDER).

Benzoin	½ lb.
Cascarilla	¼ lb.
Musk	15 grains.
Santal wood	1 lb.
Saltpetre	3½ oz.
Vetiver root	5½ oz.
Olibanum	1 lb.
Cinnamon	5½ oz.

Dissolve the saltpetre in water, saturate the powders with the solution, dry the mass, and again reduce it to powder.

This powder, strewn on a warm surface such as the top of a stove, takes fire spontaneously and gradually disappears.

FUMIGATING PAPERS AND WICKS (BRUGES RIBBONS).

French—Papier à fumigations. Ruban de Bruges; *German*— Räucherpapiere. Räucherbänder.

Fumigating papers are strips impregnated with substances which become fragrant on being heated; such a strip need merely be placed on a stove or held over a flame in order to perfume a whole room. Fumigating papers are divided into two groups: those meant to be burned, and those meant to be used repeatedly. The former, before being treated with aromatics, are dipped into saltpetre solution; the latter, in order to render them incombustible, are first dipped into a hot alum solution so that they are only charred by a strong heat, but not entirely consumed.

A. INFLAMMABLE FUMIGATING PAPER.

Papier Fumigatoire Inflammable.

The paper is dipped into a solution of 3½ to 5½ ounces of saltpetre in water; after drying it is immersed in a strong tincture of benzoin or olibanum and again dried. An excellent paper is made according to the following formula:

Benzoin	5½ oz.
Santal wood	3½ oz.
Olibanum	3½ oz.
Oil of lemon grass	150 grains.
Essence of vetiver	1¾ oz.
Alcohol	1 qt.

For use, the paper is touched with a red-hot substance, not a flame. It begins to glow at once without bursting into flame, giving off numerous sparks and a pleasant odor.

B. Non-inflammable Fumigating Paper.

Papier Fumigatoire Permanent.

This paper is prepared by dipping it in a hot solution of 3½ oz. of alum in one quart of water; after drying, it is saturated with the following mixture:

Benzoin...................................... 7 oz.
Tolu balsam.................................. 7 oz.
Tincture of tonka........................... 7 oz.
Essence of vetiver.......................... 7 oz.
Alcohol...................................... 20 fl. oz.

This paper, when heated, diffuses a very pleasant odor and can be used repeatedly. It does not burn, and strong heat only chars it. Some manufacturers make inferior fumigating papers by dipping the alum paper simply in melted benzoin or olibanum.

C. Fumigating Ribbons

are nothing but fine flat lamp wicks treated first with saltpetre solution and then with the preceding mixture. The wick is rolled up and placed in a vessel provided with a lamp burner. It is inserted in the burner like any other wick and when lighted burns down to the metal and goes out unless screwed up higher. Fumigating vessels provided with these wicks are very practical because, if artistic in form, they form quite an ornament to the room and can be instantly set in operation. A French formula gives the following mixture for saturating the wicks:

Benzoin...................................... 1 lb.
Musk... ¾ oz.
Myrrh.. 3½ oz.
Tolu balsam.................................. 3½ oz.
Tincture of orris root....................... 1 pint.
Oil of rose.................................. 15 grains.

FUMIGATING WATERS AND VINEGARS (EAUX ENCENSOIRES, VINAIGRES ENCENSOIRES).

These fluids are nothing but strong solutions of various aromatics in alcohol, a few drops of which suffice, if evaporated on a warm plate, to perfume a large room.   The following is a good formlua for fumigating water.

Benzoin................	7 oz.
Cascarilla...............................	3½ oz.
Cardamoms...........................	3½ oz.
Mace....	1¾ oz.
Musk..	150 grains.
Peru balsam................. :...........	1¾ oz.
Storax........	1¾ oz.
Tolu balsam...........................	1¾ oz.
Olibanum...........	3½ oz.
Orris root.............	14 oz.
Civet........	150 grains.
Cinnamon............................	7 oz.
Oil of bergamot......................	1½ oz.
Oil of lemon...........................	1½ oz.
Oil of geranium.......................	¾ oz.
Oil of lavender.......................	¾ oz.
Oil of neroli........	150 grains.
Alcohol............................	2 qts.

Of course, this liquid must be filtered after prolonged maceration.   By adding to it 1½ oz. of glacial acetic acid we obtain the so-called fumigating vinegar which is very useful for expelling bad odors.

FUMIGATING POWDERS (POUDRES ENCENSOIRES).

These powders which need only to be heated in order to diffuse one of the most pleasant odors, are easily prepared by intimately mixing the ground solids with the oils by means of a spatula.   We add three renowned formulas for the manufacture of such powders.

A. Poudre Impériale.

Benzoin	3½ oz.
Cascarilla	1¾ oz.
Lavender	1¾ oz.
Rose leaves	1¾ oz.
Santal wood	1¾ oz.
Olibanum	3½ oz.
Orris root	3½ oz.
Cinnamon	1¾ oz.
Oil of lemon	75 grains.
Oil of clove	30 grains.
Oil of patchouly	15 grains.

B. Poudre de la Reine.

Benzoin	7 oz.
Cedar wood	1 lb.
Cinnamon	14 oz.
Lavender	10½ oz.
Rose leaves	10½ oz.
Patchouly herb	3½ oz.
Vetiver root	3½ oz.
Civet	150 grains.
Oil of bergamot	¾ oz.
Oil of lemon	¾ oz.
Oil of neroli	150 grains.
Oil of clove	150 grains.

C. Poudre Royale.

Cinnamon	½ lb.
Cloves	½ lb.
Orris root	12½ oz.
Storax	12½ oz.
Lavender	1 lb.
Oil of clove	⅜ oz.
Oil of lavender	⅜ oz.
Oil of bergamot	⅜ oz.
Oil of lemon	⅜ oz.

CHAPTER XXI.

ANTISEPTIC AND THERAPEUTIC VALUE OF PERFUMES.

WHILE the popular use of perfumes, of course, is due to the pleasurable sensations resulting from the inhalation of their agreeable odor, it is well to call attention to the fact that they have a recognized antiseptic and therapeutic value as well.

A belief in the antiseptic value of perfumes is very ancient, and the custom of burning aromatic substances was general in times of epidemics during the middle ages, but it was not until very recently that this belief was confirmed by practical experiments.

Criton, Hippocrates, and other ancients, classed perfumes among medicines, and prescribed them for many diseases, especially those of a nervous kind. Pliny also attributes therapeutic properties to various aromatic substances, and some perfumes are still used in modern medicine.

The late Mr. W. P. Ungerer was the first modern observer to call attention to the antiseptic qualities of perfumes in general. He suggested that the fact of so few cases of tuberculosis existing in the flower-growing districts of France was attributable to the atmosphere being so full of the germ-killing odors of the flowers. He also pointed out that many working in perfume laboratories were free from disease of the respiratory organs, and that those with bronchial affections often unconsciously cured themselves in the atmosphere filled with the odors of the volatile oils.

Later this theory, based upon years of observation, was confirmed by several series of scientific experiments as to the germicidal qualities of several of the essential oils used in perfumes. M. Chamberland, of the Pasteur Institute in Paris, MM. Cadeac, Meunier, and Smetchensko experimented along the same line and M. Charrin presented a note to the French Biological Society, emphasizing and supporting their conclusions.

Tests were first made upon the germs of glanders and yellow fever, and these germs were easily killed by the odors of the essential oils. Later experiments were made with a number of oils under ordinary temperature. The oils found most effective were Ceylon Cinnamon, Chinese Cinnamon, Close, Origanum, French Geranium, Algerian Geranium, Indian Verbena, Lavender, Patchouly, Angelica, Juniper, Sandal, Bitter Orange, Cedar, Thyme, Lemon, Peppermint, French Verbena, Pine, Wormwood, and Cubeb, as well as extracts of Jasmine and Tuberose.

In order to test the action of the oils upon germs as usually encountered in the air, on walls, or on the human body, the experimental tests were made as follows:

The end of a fine platinum wire was covered with gelatin containing the culture to be tried. This wire was fixed into a cork and the cork put into the end of a test tube in the bottom of which was some of the oil being tested. At the end of a given time sterilized gelatin was pricked by the germ-bearing wire and then heated to bring about growth.

The bacteria tested in this way were: Golden Staphylococcus, Streptococcus, Coli bacilli, Tetrageni, and Bacilli Virgule. Of the five it was found that the Golden Staphylococci had the greatest power of resistance.

It was ascertained that the germs were still alive after being exposed to the vapors of the following oils for 72 hours:

Angelica, Patchouly, Lemon, Bitter Orange, Juniper, and Sandal; but were killed in that time when exposed to French Geranium, Peppermint, Origanum, Pine, and Thyme.

Sixty hours was long enough to kill the germs by Wormwood, Cedar, Cubeb, Algerian Geranium; and 48 hours was sufficient to sterilize by the volatilization from Ceylon and Chinese Cinnamon, Clove, Lavender, French and Indian Verbena, extract of Jasmine, and Tuberose.

Results were more or less contradictory when tests were made for 48 hours or less. Sometimes germs seemed to be dead after 24 hours, and in other cases the same germs were alive after 36 hours of exposure.

Further experiments showed that the Tetragene bacilli were killed in 48 hours by all the oils except Bitter Orange, Peppermint, and Cubeb; Streptococcus was killed in 48 hours by all the oils except Bitter Orange; the Virgule bacillus was made innocuous by all the oils after only 4 hours; Coli bacilli could not resist for 24 hours the vapors of Ceylon Cinnamon, Clove, French Verbena, and Tuberose. The Golden Staphylococci were also killed in 24 hours by the evaporations from Ceylon and Chinese Cinnamon, Lavender, Clove, Verbena, Jasmine, and Tuberose.

Continued experiments carried to very fine extremes went to prove that many of the bacteria were killed in less than an hour by the evaporations from the oils mentioned, and in some instances only a few minutes of exposure to these oils means death to the germs.

The importance of these investigations can hardly be exaggerated, for especially in time of epidemic the value of perfumes cannot be discounted, and even under ordinary circumstances it is known that the air is filled with germs of all kinds, which are best combated by such pleasant germicides as our perfumes.

The opinion to the contrary which is sometimes expressed, is generally based upon a misunderstanding of the subject, or is the result of imagination. It is true that flowers, if left in a closed sleeping apartment all night, will sometimes cause headache and sickness, but this proceeds not from the diffusion of their aroma, but from the carbonic acid they evolve during the night. If a perfume extracted from these flowers were left open in the same circumstances, no evil effect would arise from it.

As regards the claims of imaginative persons that they cannot tolerate various odors, all that can be said is that some delicate people may be affected by certain perfumes; but the same person to whom a musky scent would give a headache might derive much relief from a perfume with a citrine basis. Imagination has, besides, a great deal to do with the supposed noxious effects of perfumes. Dr. Coloquet, who may be deemed an authority on this subject, of which he made a special study, says in his able "Treatise on Olfaction": "We must not forget that there are many effeminate men and women to be found in the world who imagine that perfumes are injurious to them, but their example cannot be adduced as a proof of the bad effect of odors. Thus Dr. Thomas Capellini relates the story of a lady who fancied she could not bear the smell of a rose, and fainted on receiving the visit of a friend who carried one, and yet the fatal flower was only artificial."

CHAPTER XXII.

CLASSIFICATION OF ODORS.

ODORS have been classified in various ways by learned men. Linnaeus, the father of modern botanical science, divided them into seven classes, three of which only were pleasant odors, viz., the aromatic, the fragrant, and the ambrosial; but, however good his general divisions may have been, this classification was far from correct, for he placed carnation with laurel leaves, and saffron with jasmine, than which nothing can be more dissimilar. Fourcroy divided them into five series, and De Haller into three. All these were, however, more theoretical than practical, and none classified odors by their resemblance to each other. More lately Rimmel has made a new classification, comprising only pleasant odors, by adopting the principle that, as there are primary colors from which all secondary shades are composed, there are also primary odors with perfect types, and that all other aromas are connected more or less with them.

The smallest number of types to which Rimmel was able to reduce his classification is 18, and even then there are some particular odors, such as Wintergreen, which it would be difficult to introduce into any class; neither does the list comprise the scents produced by blending several different odors.   The types adopted by Rimmel are as follows:

CLASSIFICATION OF ODORS.

Classes.	Types.	Odors belonging to the same Class.
Rose	Rose	Geranium, Sweetbrier, Rhodium, Rosewood.
Jasmine	Jasmine	Lily of the Valley.
Orange Flower	Orange Flower	Acacia, Syringa, Orange leaves.
Tuberose	Tuberose	Lily, Jonquil, Narcissus, Hyacinth.
Violet	Violet	Cassie, Oris-root, Mignonette.
Balsamic	Vanilla	Balsam of Peru and Tolu, Benzoin, Styrax, Tonka Beans. Heliotrope.
Spice	Cinnamon	Cassia, Nutmeg, Mace, Pimento.
Clove	Clove	Carnation, Clove Pink.
Camphor	Camphor	Rosemary, Patchouly.
Sandal	Sandalwood	Vetivert, Cedarwood.
Citrine	Lemon	Bergamot, Orange, Cedrat, Limette.
Lavender	Lavender	Spike, Thyme, Serpolet, Marjoram.
Mint	Peppermint	Spearmint, Balm, Rue, Sage.
Aniseed	Aniseed	Badiane, Caraway, Dill, Coriander, Fennel.
Almond	Bitter Almonds	Laurels, Peach Kernels, Mirbane.
Musk	Musk	Civet, Musk-seed, Musk-plant.
Amber	Ambergris	Oak-Moss.
Fruit	Pear	Apple, Pineapple, Quince.

CHAPTER XXIII.

SOME SPECIAL PERFUMERY PRODUCTS.

Besides the preparations enumerated in the preceding pages, we find in perfumery some products which are in favor on account of their fragrance and are suitable for scenting ladies' writing-desks, sewing-baskets, boxes, and similar objects. They find their most appropriate use in places where an aromatic odor is desired, while there is no room for keeping the substances themselves. These must therefore be put into a small compass, and the aromatics chosen should be distinguished by great intensity and permanence of odor.

We subjoin a few formulas for the manufacture of such specialties, and add the remark that besides the aromatics there given other substances may be used in their preparation; but that the presence of benzoin, musk, or civet, even in small amount, is always necessary, since these substances, as above stated, not only possess an intense and permanent odor, but have the valuable property of imparting lasting qualities to more volatile odors.

It is a good plan, too, to keep on hand two kinds of these specialties—one containing musk, the other none—for the reason that the musk odor is as disagreeable to some persons as it is pleasant to others.

SPANISH SKIN (PEAU D'ESPAGNE, SPANISCH LEDER).

The article sold under this name resembles in some respects sachets or scent bags and is made as follows.

Take a piece of wash-leather (chamois), trim it to a square shape, and leave it for three or four days in the following mixture:

Benzoin.............................	½ lb.
Oil of bergamot......................	¾ oz.
Oil of lemon.........................	¾ oz.
Oil of lemon grass....................	¾ oz.
Oil of lavender......................	¾ oz.
Oil of nutmeg........................	150 grains.
Oil of clove.........................	150 grains.
Oil of neroli........................	1½ oz.
Oil of rose..........................	1½ oz.
Oil of santal........................	1½ oz.
Tincture of tonka....................	¾ oz.
Oil of cinnamon......................	150 grains.
Alcohol.............................	1 qt.

At the end of the time named remove the leather from the liquid, let it drain, spread it on a glass plate, and when dry coat it on the rough side, by means of a brush, with a paste prepared in a mortar from the following ingredients:

Benzoic acid, sublimed................	150 grains.
Musk................................	15 grains.
Civet...............................	15 grains.
Gum acacia..........................	1 oz.
Glycerin............................	¾ oz.
Water...............................	1¾ oz.

The leather is then folded in the centre, smoothed with a paper-knife, put under a weight, and allowed to dry. The dried leather forms the so-called perfume skin which retains its fine odor for years. Instead of the above alcoholic liquids any desired alcoholic perfume may be used; especially suitable are those containing oils of lemon grass, lavender, and rose, since they are not very volatile, and when combined with musk and civet remain fragrant for a long time. A sufficiently large piece of perfume skin inserted in a desk pad or placed among the paper will make the latter very fragrant. Spanish skin is chiefly used for this purpose, as well as for work, glove, and handkerchief boxes, etc. It is generally inclosed in a heavy silk cover.

If leather be thought too expensive, four to six layers of blotting-paper may be perfumed in the same way and properly inclosed. Thin layers of cotton wadding between paper can also be thus perfumed and used for filling pin-cushions, etc.

SPANISH PASTE.

Mix the following substances intimately in a porcelain mortar, and add water drop by drop until a doughy mass results.

Ambergris...............................	¾ oz.
Benzoin.................................	1½ oz.
Musk....................................	¾ oz.
Vanilla.................................	¾ oz.
Orris root..............................	¾ oz.
Cinnamon................................	¾ oz.
Oil of bergamot........................	1½ oz.
Oil of rose.............................	¾ oz.
Gum acacia..............................	1½ oz.
Glycerin................................	1½ oz.

This paste, divided into pieces about the size of a hazel-nut, is used for filling the so-called cassolettes or scent boxes which are carried in the pocket, etc., like smelling bottles. Owing to its pasty consistence this preparation can be used for perfuming jewelry (small quantities are inserted within the diamond settings), fine leather goods, belts, and other articles. It is unnecessary to lengthen the list; every practical perfumer will know what objects need perfuming.

CHAPTER XXIV.

HYGIENIC AND COSMETIC PERFUMERY.

PERFUMERY is not merely called upon to act in an æsthetic direction and gladden the senses; it has another and more important aim, that is, to aid in some respects the practice of medicine.   It is not necessary to point out that in this sense, too, it acts in an æsthetic way; for health and beauty are one and inseparable.

The field relegated to perfumery with reference to hygiene is extensive, comprising the care of the skin, the hair, and the mouth.   But we also find in commercial perfumery articles which possess no medicinal effect and serve merely for beautifying some parts of the body, for instance, paints and hair dyes.   As it is not possible to separate perfumes with hygienic effects from cosmetics, we shall describe the latter in connection with the former.

To repeat, hygienic perfumery has to deal with such substances as have really a favorable effect on health.   No one will deny that soap takes the first place among them.   Soap promotes cleanliness, and cleanliness in itself is essential to health.   But it would exceed the scope of this work were we to treat in detail of the manufacture of soap and its employment in the toilet; we must confine ourselves to some specialties exclusively made by perfumers and into the composition of which soap enters.   We do so the more readily since perfumers are but rarely in a position to make soap, and in most cases find it more advantageous to buy the raw material, that is, ordinary good soap, from the manufacturer and to perfume it.

15

Next to soap in hygienic perfumery stand the so-called emulsions and creams (crêmes) which are excellent preparations for the skin and pertain to the domain of the perfumer.

The human skin consists of three distinct parts: the deepest layer, the subcutaneous cellular tissue which gradually changes into true skin; the corium or true skin (the thickest layer); and the superficial scarf skin or epidermis which is very thin and consists largely of dead and dying cells; these are continually shed and steadily reproduced from the corium.

The skin contains various depressions, namely, the sudoriparous glands which excrete sweat; the sebaceous glands which serve the purpose of covering the skin with fat and thereby keep it soft, glossy, and supple; and lastly the hair follicles which contain the hairs, an appendage to the skin.

The main object of hygienic perfumery with reference to the skin is to keep these glandular organs in health and activity; it effects this by various remedies which, besides promoting the general health, improve the appearance of the skin.

As a special group of preparations is intended exclusively for the care of the skin, so another class is devoted to the preservation of the hair, and still another to the care of the mouth and its greatest ornament, the teeth. Accordingly the preparations belonging under this head will be divided into three groups—those for the skin, the hair, and the mouth.

CHAPTER XXV.

PREPARATIONS FOR THE CARE OF THE SKIN.

GLYCERIN.

PURE glycerin is a substance that has a powerful beautifying effect on the skin, by rendering it white, supple, soft,
and glossy; no other remedy will clear a sun-burnt skin in so
short a time as glycerin.   An excellent wash may be made by
the perfumer by mixing equal parts of thick, colorless glycerin and orange-flower water (or some other aromatic water
with fine odor), possibly giving it a rose color by the addition
of a very small amount of fuchsine.   Concentrated glycerin
must not be used as a wash, because it abstracts water from
the skin and thereby produces a sensation of heat or burning.

Besides common soap, the so-called emulsions, meals,
pastes, vegetable milks and creams are the best preparations
for the care of the skin; in perfumery they are even preferable to soap in some respects because they contain not only
substances which have a cleansing effect like any soap, scented
or not, but at the same time render the skin clearer, more
transparent, and more supple.

EMULSIONS.

Many perfumers make a definite distinction between two
groups of emulsions which they call respectively "emulsions"
and "true emulsions."   By "emulsions" they mean masses
which have the property of changing on contact with water
into a milky fluid or becoming emulsified; the term "true
emulsions" is applied to such preparations as already contain

a sufficient amount of water and therefore have a milky appearance. Hence the difference between the two preparations lies in the lesser or greater quantity of water, and is so variable that we prefer to describe them under one head.

The cause of the milky appearance of the emulsions on coming in contact with water is that they contain, besides fat, substances which possess the property of keeping the fat suspended in form of exceedingly minute droplets which make the entire fluid look like milk. As a glance through the microscope shows, the milk of animals consists of a clear fluid in which the divided fat droplets (butter) float; these by their refractive power make the milk appear white.

While soaps always contain a certain quantity of free alkali, a substance having active caustic properties, emulsions include very little if any alkali, and, since they possess the same cleansing power as soap without its disadvantages with reference to the skin, their steady use produces a warm youthful complexion, as well as smoothness and delicacy of the skin.

Glycerin is of special importance in the composition of emulsions. Besides the above-mentioned property of this substance of keeping the skin soft and supple, it acts as a true cosmetic by its solvent power of coloring matters: a skin deeply browned by exposure to the sun is most rapidly whitened by the use of glycerin alone. Moreover, glycerin prevents the decomposition of the preparations and keeps them unchanged for a long time. This quality has a value which should not be underestimated; for all emulsions are very apt to decompose and become rancid owing to the finely divided fat they contain. Under ordinary conditions, only complete protection against light and air can retard rancidity, which is accompanied by a disagreeable odor not to be masked by any perfume; an addition of glycerin, which we incorporate in all emulsions, makes them more permanent owing to the antiseptic property of this substance.

Recent years, however, have made us acquainted with a substance which in very minute quantities—one-half of one per cent of the mass to be preserved by it—prevents decomposition and rancidity of fats. This is salicylic acid, a chemical product which, being harmless, tasteless, and odorless, should be employed wherever we wish to guard against destructive influences exerted by air, fermentation, etc. While formerly all emulsions were made only in small amounts, just sufficient for several weeks' use, salicylic acid enables us to manufacture larger quantities at once and to keep them without much fear of their spoiling. However, even the presence of salicylic acid is no guaranty against deterioration, if other precautions are neglected. The products should be kept in well-stoppered bottles or vessels, in a cool and dark place. All substances cannot be preserved by salicylic acid, and there are certain ferments or fungi which resist the action of salicylic acid. If chloroform is not objectionable in any of these preparations— and only so much is necessary as can be held in actual *solution* by the liquid, on an average three drops to the ounce—this preservative is preferable to salicylic acid.

The fats most used in the preparation of emulsions are expressed oil of almonds, olive oil, and lard. Almond oil is best made by immediate pressure of the bruised fruits, since fresh almond meal likewise finds application in perfumery; olive oil and lard must be very carefully purified. This is done by heating them for one hour with about ten times the quantity of water containing soap (one per cent of the quantity of fat to be purified). They are then treated five or six times with pure warm water until the latter escapes quite neutral. If the water turns red litmus paper blue, it would indicate the presence of free alkali (soap); if it turns blue litmus paper red, it would prove the presence of free fatty acids (rancid fat). Either one of these substances, especially the latter, would injure the quality of the product. The fat should be absolutely neutral and have no influence on either kind of litmus paper; then its quality may be pronounced perfect.

CHAPTER XXVI.

MANUFACTURE OF CASEIN.

MOST massage or rolling creams are composed mainly of casein, which is the principal albuminoid matter of milk and is extracted from skimmed milk. For this purpose a solution of concentrated sulphate of magnesium is added to the milk; compact flakes are precipitated and dissolved again in pure water; the product is filtered and the casein precipitated by acetic acid.

Thus obtained, this matter is of a whitish or yellowish color, insoluble in water and alcohol, and soluble in alkalies and alkali carbonates. It is the presence of these latter in the milk that allows the casein to remain for the most part dissolved.

After centrifugation the milk is placed in large vats where the casein is coagulated, either with rennet or an acid or, more simply, by allowing the coagulation to form of itself. Casein coagulated by rennet is precipitated by sulphuric acid; that coagulated with an acid is precipitated by hydrochloric acid; that allowed to coagulate spontaneously, by lactic fermentation, is left as it is.

According to the uses to which it is to be put, casein undergoes mechanical and physical treatments with a view of transforming it into powder, or thoroughly purifying and drying it to facilitate its preservation.

CASEIN BY ELECTROLYSIS.

Casein has been obtained by electrolysis, as follows: In the middle of a large vat full of skimmed milk heated to a temperature of 80° C., a porous vessel is placed containing a 5 per cent solution of caustic soda; an iron cathode is plunged into the soda and a rod of carbon, serving as anode, into the milk. The two electrodes are 3 inches apart. By sending through the apparatus a current of a density of 1 ampere per square centimetre (0.155 square inch) of the anode, the phosphoric acid contained in the milk is entirely set free, and the casein precipitates. With a current of 160 amperes under 11 volts it is possible, in 20 minutes only, to coagulate completely the casein from 100 litres (26.42 gallons) of skimmed milk. The caustic soda can be replaced by an equal volume of skimmed milk derived from a preceding operation, the electrodes being disposed in the same manner. With a current of 160 amperes under 18 volts, 10 minutes are sufficient to precipitate the casein of 100 litres (26 gallons) of skimmed milk. This method, due to M. Gateau, presents three important advantages:

The production is greater because the weight of casein thus obtained is superior to that obtained by acids or rennet.

The cost price is very low; the cost of the electric power is less than that of acids or rennet, and decreases especially when the current is produced by water or wind power.

Casein thus obtained contains no impurities; foreign elements cannot penetrate the liquid, as the precipitation is effected by the electric current alone.

VEGETABLE CASEIN.

Vegetable casein, which can be put to the same uses as animal casein, has been extracted on a large scale from the soya bean, which has the following composition: Water, 12.35 per cent; dust and other impurities, 7.90 per cent; dry pods, 7.85 per cent; dry casein, 25.55 per cent; residues, 29.80 per cent; oils and fatty substances, 16.42 per cent.

The beans are first washed to rid them of all dust and foreign substances, then dried lightly in the open air. They are then triturated to remove the fatty substances they contain. This is done with the aid of crushers and presses such as are used for the extraction of olive oil. Following this the pulp is triturated between millstones with ordinary water, and a milky liquid is thus obtained that may be concentrated by repeating the operation several times on the same pulp. It is then clarified by means of appropriate filter presses. To separate the casein from this liquid it is heated by means of an alembic placed in vessels furnished with agitators; coagulation takes place.

CHAPTER XXVII.

FORMULAS FOR EMULSIONS.
AMANDINE.

Almond Cream.—Melt ten pounds of purified lard in an enamelled iron pot or a porcelain vessel, and while increasing the temperature add little by little five pounds of potash lye of 25% strength, stirring all the time with a broad spatula. When fat and lye have become a uniform mass, 2¾ to 3½ ounces of alcohol is gradually added, whereby the mixture acquires a translucent, crystalline appearance. Before the alcohol is added three-fourths to one ounce of oil of bitter almond is dissolved in it. The soapy mass thus obtained is called " almond cream " (crême d'amandes) and may be used alone for washing. For making Amandine take of—

Expressed oil of almonds	10 lb.
Almond cream	3½ oz.
Oil of bergamot	1 oz.
Oil of bitter almond	1½ oz.
Oil of lemon	150 grains.
Oil of clove	150 grains.
Oil of mace	150 grains.
Water	1¾ oz.
Sugar	3½ oz.

In the manufacture the following rules should be observed.

Effect the mixture in a cool room, the cellar in summer, a fireless room in winter. Mix the ingredients in a shallow, smooth vessel, best a large porcelain dish, using a very broad, flat stirrer with several holes. The sugar is first dissolved in the water and intimately mixed with the almond cream. The essential oils are dissolved in the almond oil contained in a vessel provided with a stopcock. The oil is first allowed to run into the dish in a moderate stream under continual stir-ring. The mass soon grows more viscid, and toward the end of the operation the flow of oil must be carefully restricted so that the quantity admitted can be at once completely mixed with the contents of the dish. Well-made amandine must be rather consistent and white, and should not be translucent. If translucency or an oily appearance is observed during the mixture, the flow of oil must be at once checked or enough almond cream must be added to restore the white appearance, under active stirring.

As amandine is very liable to decompose, it must be im-mediately filled into the vessels in which it is to be kept, and the latter, closed air-tight, should be preserved in a cool place. By adding ¾ ounce of salicylic acid, amandine may be made quite permanent so that it can be kept unchanged even in a warm place

We have described the preparation of amandine at greater length because its manufacture requires some technical skill and because the preparation of all other cold-creams corre-sponds in general with that of amandine.

GLYCERIN EMULSIONS. A. GLYCERIN CREAM.

Glycerin	½ lb.
Almond oil	14 oz.
Rose water	12½ oz.
Spermaceti	3½ oz.
Wax	480 grains.
Oil of rose	60 grains.

Melt the wax and spermaceti by gentle heat, then add the almond oil, next the glycerin mixed with the rose water, and

lastly the oil of rose which may also be replaced by some
other fragrant oil or mixture. If the preparation is to be
used in summer, it is advisable to increase the wax by one-
half, thus giving the mass greater consistence.

B. Glycerin Jelly.

Glycerin............................. 2 lb.
Almond oil........................... 6 lb.
Soap......... .... 5½ oz.
Oil of orange peel......................150 grains.
Oil of thyme.......................... ¾ oz.

Mix the soap with the glycerin, gradually add the oil (as
for amandine), and finally the aromatics.

CHAPTER XXVIII.

FORMULAS FOR CREAMS.

Massage Creams.

The various substances used for the basis of massage creams are casein, tragacanth, starch, Irish moss, quince mucilage, gelatin, and isinglass. All these give non-greasy preparations, but frequently a little oil is added to assist in lubrication.

Casein is the proteid substance of milk, which is prepared by adding acetic acid to separated milk. The casein is collected on a calico filter, pressed, and dried at a temperature not exceeding 40° C. Casein is soluble in alkalies and in alkaline carbonates, and in borax solution with which it forms a jelly. Usually a small quantity of fat is added, such as cacao butter or liquefied paraffin, which serves as a lubricant to the skin. The following formula will give a suitable cream:

Casein Cream.

Casein...............................	2 oz.
Borax................................	½ oz.
Liquid paraffin.......................	½ oz.
Boric acid............................	½ oz.
Glycerin..............................	1 oz.
Water................................	12 oz.

Dissolve the borax in the water and rub with the casein to a cream in a warm mortar. Dissolve the boric acid in glycerin and mix with the casein solution. Then add the liquid paraffin. If desired it may be tinted with a small quantity of carmine or an aniline dye and perfumed as required.

Casein Rolling Cream.

Dr. Brewer's Formula.

Sweet milk (skimmed or normal)............	15 gallons.
Solution of formaldehyde..................	1 oz. 7 drs.
Boiling water..........................	4 gallons.
Borax.................................	1 lb. 14 oz.
Alum..................................	3 lbs. 12 oz.
Boric acid.............................	7 lbs.
Cacao butter...........................	1 lb. 10 oz.
Hydrous wool-fat.......................	1 lb. 10 oz.
Solution of carmine.....................	2 oz. 2 drs.
Oil of rose geranium.....................135 minims.
Oil of bitter almond.....................	45 minims.
Water (for washing).....................	15 gallons.

Add the solution of formaldehyde to the milk; mix well; add the solution of carmine; stir, and heat to 50° C. Dissolve the borax in 2 gallons of boiling water; add this to the milk mixture; stir quickly; heat to 50° C.; and strain through muslin. Dissolve the alum in the rest of the boiling water; add slowly to the other liquid, stirring constantly. Allow the curd to settle; drain off the liquid; wash the curd with the water; squeeze off the moisture until the curd weighs 25 pounds. Melt the cacao butter and mix it with the wool-fat; add to this the boric acid and mix thoroughly. Incorporate the fatty mass with the curd; add the perfume; mix thoroughly; and fill into jars with air-tight covers.

Massage Rolling Cream.

Sweet milk.............................	1 gallon.
Borax.................................	3 oz.
Boracic acid...........................	3 oz.
Soda benzoate..........................	½ oz.
Tartaric acid...........................	6 oz.

Put the borax, boracic acid, and benzoate soda in the milk and heat to the boiling point, then remove from the fire and add the tartaric acid; let it stand 24 hours, then strain and add glycerin 1½ ounces to the residue. Color and perfume to suit.

TRAGACANTH CREAM.

Tragacanth in powder	2 oz.
Glycerin	8 oz.
Boracic acid	1 oz.
Powdered borax	1 oz.
Alcohol	5 oz.
Distilled water	48 oz.
Tincture of benzoin	1 oz.
Oil of bergamot	60 minims.
Oil of orange flowers	30 minims.
Oil of rose geranium	60 minims.

Soak the tragacanth in the water for 24 hours in a wide-mouthed bottle, occasionally agitating it. Add the glycerin, cut the oils in the alcohol and then gradually add, with constant stirring, when a thick cream results. Add the tincture of benzoin and thoroughly incorporate.

CHONDRUS CREAM.

Irish moss (chondrus)	1 oz.
Distilled water	10 oz.
Glycerin	2 oz.
Boracic acid	1 dram.
Eau de cologne	1 oz.

Wash the Irish moss (which should be as white as possible) in a little cold water, then boil gently for 10 minutes in the 10 ounces of water in a closed vessel, and strain through muslin. Add the glycerin and the borax, previously dissolved in 1 ounce of water, cool, and finally add the eau de cologne.

Isinglass Jelly.

Isinglass	1 oz.
Water	24 oz.
Glycerin	6 oz.
Boric acid	2 drams.

Soak the isinglass in the water over night, then gently heat until dissolved. While still hot add the water and glycerin in which the boric acid has been previously dissolved, and allow to stand until cold. Gelatin may be used in place of isinglass, but will not make a clear mixture. Perfume to suit.

Frozen Foam Cream.

Agar-agar	1 oz.
Water	8 oz.
Stearic acid	1 oz.
Sodium carbonate	5½ drams.
Oil of theobroma	1 oz.
Alcohol	2½ drams.

Agar-agar, or Japanese isinglass, is a gelatinous substance prepared from a species of seaweed. It is not as soluble as isinglass but dissolves in boiling water, forming a transparent jelly when cold.

Dissolve the agar-agar in 5 ounces of boiling water and strain. To 3 ounces of water in a water-bath add the stearic acid and the sodium carbonate; when action ceases add the theobroma and agar-agar. Mix thoroughly by means of an egg-beater; then remove the dish from the water bath and continue agitating until a uniformly smooth lather, measuring about three times the volume of the contained liquid, results. When nearly cold add the perfumes desired.

GLYCERIN JELLY.

Gelatin	1 oz.
Water	24 oz.
Glycerin	12 oz.
Boro-glycerate	12 oz.

Soak the gelatin over night in the water, then dissolve by gentle heat and add the other ingredients. Pour into wide-mouthed glass jars while still warm and allow to cool.

FACE CREAMS.

The most popular face creams of the present day are the so-called "Greaseless" preparations, which are sold under various fancy names, such as "Vanishing Cream," etc. Their great advantage is their faculty of disappearing when rubbed into the skin, leaving no greasiness behind. They have also cleansing properties and leave the skin clean and soft, with a natural bloom, and, moreover, do not choke the pores, as they are practically devoid of mineral matter. A typical cream of this kind is made thus:

VANISHING CREAM.

Dr. J. S. Brewer's Formula.

Stearic acid (white, triple-pressed)	4 lbs. 12 oz.
Glycerin	8 lbs. 8 oz.
Distilled water	14 pints.
Stronger ammonia water	4 oz. 6 drs.
Cologne spirits	1 pint.
Oil of hyacinth	6 drops.
Oil of jasmine (artificial)	4 drams.
Artificial musk (crystal)	20 grains.
Terpineol	2 oz.

Melt the stearic acid on a water bath at 75° to 80° C. Heat 2 pounds of glycerin with 12 pints of water to the same temperature; add the ammonia water; and pour slowly into the melted stearic acid, with constant stirring. Mix the rest of the glycerin and water and heat to 80° C.; pour this into the first mixture, with constant stirring; continue the temperature and the stirring for about fifteen minutes. Remove from the heat and beat until cold. Mix the perfuming materials with the spirits and add this slowly, with constant beating, to the cream.

COOBAN'S FORMULA.

Stearic acid	2 oz., av.
Sodium carbonate	1¼ oz., av.
Powdered borax	¼ oz., av.
Glycerin	4 fl. oz.
Water	32 fl. oz.
Oil of ylang-ylang	80 drops.
Oil of rose	20 drops.
Heliotropin	20 grains.
Alcohol	4 fl. oz.

Heat the first five ingredients together on a water bath until effervescence ceases; then remove and stir at intervals until the mixture begins to stiffen. Incorporate the heliotropin and the oils dissolved in the alcohol and beat up the whole with a paddle or egg-beater. Upon reheating and beating again a fluffy, creamy product is obtained.

STANISLAUS'S FORMULA.

Stearic acid	1½ oz., av.
Cacao butter	¼ oz., av.
Sodium carbonate, pure	1 oz., av.
Borax, powdered	¼ oz., av.
Glycerin	1¼ fl. oz.
Water	20 fl. oz.
Mucilage of tragacanth	5 fl. oz.
Terpineol	45 minims.
Oil of bitter almond	2 drops.
Oil of rose	15 drops.
Alcohol	1½ fl. oz.

Dissolve the borax and sodium carbonate in the water and add this solution, with the glycerin and mucilage to the cacao butter and stearic acid contained in a vessel on a water-bath. Heat the whole together until effervescence ceases, allow to cool, then add the perfumes dissolved in the alcohol and beat until the mass stiffens.   Heat again and beat until it becomes creamy and fluffy.

WITCH HAZEL CREAM.

Stearic acid.............................. 4 oz., av.
Sodium carbonate......................... ½ oz., av.
Glycerin................................. 4 fl. drs.
Water.................................... 16 fl. oz.
Distilled extract of witch hazel............. 20 fl. oz.

Dissolve the sodium carbonate in the water and add to the glycerin contained in a large evaporating dish.   Then add the stearic acid and heat the mixture on the water-bath until effervescence has ceased and a clear solution results.   Keep this near the boiling point for at least an hour, stirring frequently and making up for loss of water through evaporation by the addition of more water, being careful not to add too much.   Now add the witch hazel extract, transfer the whole to a hot mortar or other suitable dish and beat until it is of the proper consistency.

GREASELESS COLD CREAM—A SIMPLE METHOD.

If it is desired to make a small quantity of cream without the use of a water-bath or grinding mill, it may be done as follows:

Fill the wooden bucket of an ice-cream freezer with scalding hot water; in the tin bucket place 4 ounces stearic acid, ½ ounce paraffin, and 12 ounces glycerin; melt these all together by placing the tin bucket in the hot water.

Add to this mixture when dissolved ½ ounce stronger ammonia (26 deg.), and, turning the crank, stir for about ten minutes, or until a perfect saponification results. Then take 16 ounces warm distilled water, in which have been previously dissolved 15 grains powdered borax, add to the other mixture, and stir again, thoroughly mixing all together.

Allow this to stand about twenty-four hours so it will drop, then add 100 minims of lilac oil, and mix again. The cream is then ready to box. In warm weather it may be necessary to add more paraffin to make the cream stiffer. This makes a most excellent cream and may be guaranteed not to dry out.

DISAPPEARING CLEANSING CREAM.

Boiling water	48 oz.
Quince seed, clean	2 oz.

Pour the boiling water on the quince seed. Let it stand until cold, with occasional stirring; then strain well and dissolve in the liquid one-quarter ounce of benzoate of soda. Then heat the liquid, being careful not to scorch it, and dissolve in it 4 pounds of stearic acid, finely ground.

Place 6 pints of water in a large graniteware vessel and dissolve in it 2 ounces of dried carbonate of soda, U. S. P., and 1½ ounces powdered borax and bring to a boil, then add to this, constantly stirring, the quince-seed mixture.

When effervescence ceases, remove from the heat and stir until it thickens; then add a mixture of warm water, 120° F., 1 gallon; glycerin, 4 ounces; peroxide of hydrogen, ½ pint. If the addition of the latter mixture should cause lumpiness, return to the heat, stirring until it becomes smooth.

COLD CREAMS.

The old-fashioned cold creams are still preferred by many persons who have a dry skin, which needs lubricating in order to make up for the deficiency of natural fat. The following are some of the best formulas for a cream of this type:

COLD CREAM I.

Mineral oil...........................	1 gallon.
Paraffin..............................	1 lb.
White wax............................	3 lb.
Borax, powdered......................	3 oz.
Lukewarm water.......................	1 gallon.

Melt the wax and paraffin in the mineral oil; dissolve the borax in lukewarm water, and then pour the two together; stirring briskly less than a minute; then perfume with rose oil, using 7 drams.

For witch-hazel cream use same formula as above, using witch hazel instead of water, and leave out the perfume oil.

COLD CREAM II.

Spermaceti...........................	1 oz.
White wax............................	1 oz.
Petrolatum...........................	8 oz.
Rose water...........................	2 oz.
Benzoic acid.........................	30 grains.

Melt the spermaceti and wax over a water-bath and add the petrolatum. Dissolve the benzoic acid in the rose water by gently warming and mix with the oil and melted fats in a warm mortar, stirring vigorously until cold. The cream may be tinted if desired with solution of carmine or alkanet, and can be perfumed with oil of almonds, otto of roses, or heliotropin.

The color should be dissolved in the rose water, while the perfumes should be added last of all.

ALMOND CREAM.

Spermaceti.............................	40 grains.
White wax..............................	40 grains.
Oil sweet almonds......................	1 oz.
Powdered white Castile soap	75 grains.
Quince mucilage........................	2 oz.
Powdered borax........................	40 grains.
Alcohol................................	1 oz.
Distilled water to make................	1 pint.

Quince mucilage is made by taking 1 ounce quince seed and 32 ounces hot distilled water. Stir frequently for twenty-four hours and then strain through cheesecloth.

Melt the spermaceti, white wax, and almond oil in a double boiler; dissolve the borax and Castile soap in 2 ounces of hot water and beat gradually in with the waxes; add next the quince mucilage and alcohol, then the balance of the water, and keep on beating until smooth.

THEATRICAL COLD CREAM.

Cacao butter...........................	4 oz.
Petrolatum, white......................	1 oz.
Almond oil.............................	3 oz.
Lanolin................................	1 dram.
White wax..............................	2 drams.
Spermaceti.............................	2 drams.
Sodium borate.........................	1 dram.
Rose water.............................	8 oz.
Tincture benzoin.......................	2 drams.

Melt the cacao butter, petrolatum, wax, spermaceti, and lanolin over a water-bath; dissolve the sodium borate in the rose water, and add gradually with constant stirring. Add the benzoin, a few drops at a time, and beat smooth.

MENTHOLATED CREAM.

White petrolatum	9 oz.
White wax	1 oz.
Menthol	60 grains.
Camphor	60 grains.
Thymol	30 grains.
Boric acid	60 grains.

Melt the wax and then add the petrolatum. When the whole is fluid, remove from the heat and stir in the other ingredients. Stir until the mixture stiffens and then run it through an ointment mill or work it smooth in a mortar. It may be colored green if desired. By omitting the menthol and thymol it may be put up as a camphor cream.

WITCH-HAZEL COLD CREAM.

White wax	1 oz.
Spermaceti	1 oz.
Petrolatum	4 oz.

Melt, pour into a mortar which has been heated by being immersed some time in boiling water. Very gradually add 3 ounces of rose water and 1 ounce of witch hazel, and assiduously stir the mixture until an emulsion is formed, and afterward until the mixture is nearly cold.

CAMPHOR ICE.

Camphor ice is a popular dressing for roughness of skin. It may be prepared thus:

White petrolatum	8 oz.
Paraffin wax	2 oz.
Camphor	2 oz.

Melt the paraffin wax, add the petrolatum and lastly the camphor. Pour into jars or molds while still melted. Add more paraffin to make stiffer if desired. Color and perfume to suit.

GLYCERINATED CAMPHOR ICE.

Powdered camphor	20 grams.
Liquid petrolatum	20 grams.
Paraffin	50 grams.
Petrolatum	80 grams.
Glycerin	20 grams.
Alkanet	5 grams.

Mix the two petrolatums and the paraffin with the aid of a gentle heat; digest the alkanet in the mixture on a sand-bath for several hours. Dissolve the camphor in the heated mixture; add the glycerin; strain; and stir until cold.

CHAPTER XXIX.

FORMULAS FOR MEALS, PASTES AND VEGETABLE MILK.

MEALS AND PASTES.

The so-called meals (farines) and pastes (pâtes) really consist of the flour of fatty vegetable substances which possess the property of forming an emulsion with water and are frequently used in washes. As they are free from alkali, they are the most delicate preparations of the kind and are especially suitable for washing the face or sensitive hands.

SIMPLE ALMOND PASTE (PÂTE D'AMANDES SIMPLE).

Bitter almonds	6 lb.
Alcohol	2 qts.
Rose water	4 qts.
Oil of bergamot	10½ oz.
Oil of lemon	3½ oz.

Put the bitter almonds in a sieve, dip them for a few seconds in boiling water, when they can be easily deprived of their brown skin; carefully bruise them in a mortar, and place them in a glazed pot set in another kept full with boiling water; pour over them two quarts of the rose water heated to near the boiling-point. Keep up the heat under continual stirring until the almond meal and rose water form a uniform mass free from granules; in other words, until the meal is changed into paste. The pot is now allowed to cool somewhat, when the rest of the rose water and the oils dissolved in alcohol are added. Almond paste should have a uniform, butter-like consistence if the first part of the operation has been carefully performed.

ALMOND AND HONEY PASTE (PÂTE D'AMANDES AU MIEL).

Bitter almonds...............................	2 lb.
Yolk of egg	30 yolks.
Honey.......................................	4 lb.
Expressed oil of almond..................	4 lb.
Oil of bergamot...........................	1 oz.
Oil of lemon...............................	¾ oz.
Oil of clove	¾ oz.

Decorticate and bruise the bitter almonds and add them with the essential oils to the mixed yolks, honey, and almond oil.

ALMOND MEAL (FARINE D'AMANDES).

Almond meal...............................	4 lb.
Orris root, powdered	5½ oz.
Oil of lemon...............................	1 oz.
Oil of bitter almond	150 grains.
Oil of lemon grass.........................	75 grains.

Almond meal here means the bran left after expressing the oil from sweet almonds. First mix the powdered orris root intimately with the essential oils and triturate the mass

with the almond bran. Other essential oils may also be
used for perfuming the mass.

·PISTACHIO MEAL (FARINE DE PISTACHES).

Pistachio nuts..........................	4 lb.
Orris root, powdered...................	4 lb.
Oil of lemon...........................	1¾ oz.
Oil of neroli...........................150 grains.
Oil of orange peel.....................	1 oz.

The pistachio nuts are blanched in the same manner as
almonds (see under Simple Almond Paste), and then reduced
to a meal.

VEGETABLE MILK.

The several varieties of vegetable milk are merely emulsions containing sufficient water to give them a milky appearance. They are used as such for washes and are in great
favor. Owing to the larger amount of water they contain,
they are more liable to decompose than the preparations described above, since the fats present in them easily become
rancid on account of their fine division in the milk.

In order to render these preparations more stable, they
receive an addition of about five to ten per cent of their
weight of pure glycerin which enhances their cosmetic effect.
The addition of about one-half of one per cent of salicylic
acid is likewise to be recommended, as it makes them more
stable.

In the following pages we shall describe only the most important of these preparations usually made by the perfumer.
In this connection we may state that by slightly modifying
the substances used to perfume them, new varieties of vegetable milk can be easily prepared.

Every vegetable milk consists in the main of a base of
soap, wax, and spermaceti, and an aromatic water which gives

the name to the preparation. This composition is intended to keep suspended the fatty vegetable substances (almond or pistachio meal, etc.), thus producing a milky appearance.

Vegetable milks are made as follows.

Melt the soap with the wax and spermaceti at a gentle heat. Prepare a milk from the vegetable substance and the aromatic water (*e.g.*, *unexpressed* almonds and rose water) by careful trituration, strain it through fine silk gauze into the vessel containing the melted mixture of soap, wax, and spermaceti, stir thoroughly, let it cool, and add the alcohol holding in solution the essential oils, the glycerin (and the salicylic acid), under continual stirring. The alcohol must be added in a very thin stream, otherwise a portion of the mass will curdle. The coarser particles contained in the milk must be allowed to settle by leaving the preparation at rest for twenty-four hours, when the milk can be carefully decanted from the sediment and filled into bottles for sale.

LILAC MILK (LAIT DE LILAS).

Soap	2¼ oz.
Wax.....................................	2¼ oz.
Spermaceti..............................	2¼ oz.
Sweet almonds..........................	1 lb.
Lilac-flower water......................	4½ pints.
Huile antique de lilas...................	2½ oz.
Alcohol (80–85% Tralles)	2 lb.

In place of lilac-flower water and huile antique de lilas, lilacin (terpineol) may be used, a sufficient quantity (about 1 oz.) being dissolved in the alcohol. But the lilacin must be pure and of clean odor.

VIRGINAL MILK (LAIT VIRGINAL).

This preparation differs from all other milks sold in perfumery in that it consists of some aromatic water with tincture of benzoin and tolu. In making it, pour the aromatic water in a very thin stream into the tincture under vigorous stirring. If the water flows in too rapidly, the resins present in the tincture separate in lumps; but if slowly poured in, the resins form minute spheres which remain suspended. The preparation is named after the aromatic water it contains: Lait virginal de la rose, à fleurs d'oranges, etc. Its formula is:

Tincture of benzoin.....................	2 oz.
Tincture of tolu........................	2¾ oz.
Aromatic water	4 qts.

CUCUMBER MILK (LAIT DE CONCOMBRES).

Soap................................	1 oz.
Olive oil.............................	1 oz.
Wax.................................	1 oz.
Spermaceti...........................	1 oz.
Sweet almonds........................	1 lb.
Cucumber juice (freshly expressed).......	4½ pints.
Extract of cucumber...................	1 pint.
Alcohol..............................	2 lb.

DANDELION MILK.

Soap	2¼ oz.
Olive oil	2¼ oz.
Wax	2¼ oz.
Sweet almonds........................	1 lb.
Extract of tuberose...................	1 lb.
Rose water...........................	5 pints.
Dandelion juice................... ...	5 oz.

Dandelion juice is the bitter milk sap of the root of the common dandelion (Leontodon taraxacum); it should be expressed immediately before use. The rose water may be replaced by some other aromatic water or even ordinary water; but the latter should be distilled, otherwise the lime it contains would form an insoluble combination with the soap.

BITTER-ALMOND MILK (LAIT D'AMANDES AMÈRES).

Bitter almonds........................	2¼ oz.
Soap...............................	2¼ oz.
Expressed oil of almond........... ...	2¼ oz.
Wax	2¼ oz.
Spermaceti..........................	2¼ oz.
Rose water..........................	4 qts.
Alcohol.............................	3 pints.
Oil of bitter almond	½ oz.
Oil of bergamot.....................	1 oz.
Oil of lemon....................... ...	½ oz.

ROSE MILK (LAIT DE ROSES).

Olive oil............................	2¼ oz.
Soap	2¼ oz.
Wax	2¼ oz.
Spermaceti	2¼ oz.
Sweet almonds.......................	4 lb.
Oil of rose	150 grains.
Rose water..........................	4 qts.
Alcohol.............................	1 pint.

PISTACHIO MILK (LAIT DE PISTACHES).

Soap	2¼ oz.
Olive oil	2¼ oz.
Wax	2¼ oz.
Spermaceti.........................	2¼ oz.
Pistachio nuts.......................	14 oz.
Oil of neroli........................	¾ oz.
Orange-flower water.................	6 qts.
Alcohol	1 qt.

CHAPTER XXX.

PREPARATIONS USED UPON THE HAIR.

Cantharides.—This is the great scalp stimulant. It is the active constituent of almost every advertised remedy for baldness, and is of undoubted value in suitable cases. When of proper strength no danger attends its external use in healthy persons. But it is liable to be absorbed into the system, and is then dangerous to those who are affected by kidney disease.

Jaborandi.—This is prepared from the leaves of pilocarpus jaborandi. Its active principle, hence named Pilocarpine, is a very powerful and sometimes poisonous drug. Used externally, it promotes the growth of the hair, especially where this is dry. It has also a darkening effect upon the hair, and while therefore advantageous to those who are turning grey, its use should be avoided by those whose hair is of a naturally golden hue. It has some germicidal properties.

Resorcin.—This is said to lessen the formation of dandruff and useful when the scalp is in a very open and relaxed condition, but great attention must be paid to the strength of the preparation, for it has been shown that while a 1 to 3 per cent solution hardens the skin, a 10 to 50 per cent solution destroys it. Moreover, its continued use will stain the hair a greenish tinge. For these reasons it should not be used in solutions, but only in a salve or pomade containing not over 10 per cent.

Formalin (Schering's).—This preparation is a 40 per cent solution of formic aldehyde, and is one of the most powerful bactericides known. No bacillus resists a 3 per cent. solution. Its action upon fungi is not so marked, and if employed for ringworm a stronger solution will be necessary. Soap can be obtained containing it.

Thymol.—A more powerful antiseptic than either carbolic or salicylic acid.

Bay Rum.—This should consist of the essential oil of myrcia acris dissolved in rum. It is not probable that there is any virtue in the rum, and a pleasanter compound will result if alcohol is substituted for it. It acts as a slight stimulant to the scalp, and is a safe wash for the hair.

Quillaia, or Panama Bark.—Quillaia is obtained from quillaia saponaria. It yields a natural soap, and a little of the bark shaken up with water gives a considerable lather. It does not seem to soften the water much, but it has the great merit of not only containing no free alkali, but in being very slightly acid. It is a gentle scalp stimulant, and helps to remove dandruff.

Castor Oil.—This is frequently used in combination with bay rum, for, used alone, it mats the hair. It is the only oil that will dissolve in alcohol (1 in 3½), and mixed with eau de cologne makes a clear brilliantine. It has the merit of being absolutely safe, for it is hard to ignite it even with a naked light.

Rosemary.—This is a very ancient constituent of hair washes. There is no objection to its use.

Ammonia, Borax, and Soap.—These require to be used with care, because they leave the hair dry and brittle. Borax makes the hair lighter in color, and soap clogs the follicles. Quillaia has but weak detergent properties. Provided no strain whatever, even that of brushing, be put on the hair until dry and the smell has gone off, weak ammonia would appear to be the safest. Being volatile, it has the great advantages that, once evaporated, no further effect can take place; while with the fixed alkalies the deleterious effect may go on indefinitely.

Benzol Chemically Pure.—(Boiling point 180° C.) This is used for removing dandruff and seborrhœic crusts. It is, of course, dangerous to use near a naked light; but the kind mentioned above, which has a lower boiling-point than alcohol, is practically quite safe otherwise. Common benzol should be used only in the open air. In consequence of its drying effect, it should be followed by the application of an oily dressing.

CHAPTER XXXI.

FORMULAS FOR HAIR TONICS AND HAIR RESTORERS.

A. Simple Hair Tonics.

These are used when there is a tendency for the hair to fall out, and to prevent the hair splitting at the ends. It is doubtful whether quinin has any particular action in this way, but tincture of jaborandi (which contains pilocarpine) and tincture of cantharides both have stimulating and strengthening effects on the roots of the hair, as already noted. The following are typical tonic lotions for general use:

GLYCERIN AND CANTHARIDES TONIC.

Ammonia water...............................	3½ oz.
Tincture of cantharides (see below)..........	3½ oz.
Rosemary water.............................	8 qts.
Glycerin...................................	10½ oz.
Oil of roses...............................	¾ oz.

The tincture of cantharides is made by macerating 1¾ ounces of powdered Spanish flies (Lyetta vesicatoria) in one quart of strong alcohol. The caustic ammonia has a cleansing effect, the glycerin makes the hair soft and the entire preparation cleans and softens the hair at the same time.

JABORANDI AND CANTHARIDINE HAIR TONIC.

Tincture of jaborandi......................	1 oz.
Tincture of cantharides....................	1 oz.
Tincture of capsicum......................	1 dram.
Aromatic vinegar..........................	1 oz.
Rose water to make........................	10 fl. oz.

QUININ AND BAY RUM TONIC.

Quinin bi-sulphate	1 dram.
Tincture of cantharides	1 dram.
Alcohol	2 oz.
Bay rum to make	10 fl. oz.

Dissolve the quinin in the alcohol and add the other ingredients. This compound will have the odor and color of bay rum.

A quinin tonic with more body and a flower odor may be made as follows:

QUININ HAIR TONIC.

Sulph. quinin	60 grains.
Alcohol	2 pints.
Tincture cantharides	2 oz.
Glycerin	2 oz.
Lilac water	16 oz.
Powdered borax	35 grains.
Water to make	1 gallon.

This will have a lilac odor, but any other perfume may be substituted by omitting the lilac water. It may be colored to suit.

Neither of these two preparations will possess the so-called eau de quinin odor. A formula for the latter follows:

EAU DE QUININ TONIC.

Sulph. quinin	60 grains.
Tincture cantharides	2 oz.
Glycerin	1 oz.
Eau de quinin oil	1 oz.
Powdered borax	35 grains.
Alcohol	4 pints.
Water	4 pints.

Eau de quinin red color, 80 min., or to suit.

Dissolve the quinin and oil in the alcohol and the borax in the water; add the other ingredients; shake, and allow to stand three or four days before filtering through double filter paper. Use no magnesia.

SAGE HAIR TONIC.

Sage leaves	3 drams.
Cinchonin sulphate	20 grains.
Tincture of cantharides	120 minims.
Tincture of capsicum	120 minims.
Menthol	8 grains.
Glycerin	120 minims.
Alcohol	1 oz.
Bay rum	4 oz.
Water enough to make	16 fl. oz.

Make an infusion of the sage and strain. Dissolve the cinchonin sulphate and the menthol in the alcohol, bay rum, and tinctures. Mix all together and filter clear.

SALICYLIC HAIR TONIC.

Salicylic acid	30 grains.
Boracic acid	60 grains.
Tincture of cantharides	1 fl. oz.
Rosemary water and rose water, equal parts, to make	10 fl. oz.

Tint yellow with strong tincture of turmeric, or violet with a drop of tincture of alkanet.

ANTISEPTIC HAIR TONIC.

Carbolic acid	2 grams.
Tincture of nux vomica	7 grams.
Tincture of red Peruvian bark	30 grams.
Tincture of cantharides	2 grams.
Eau de cologne	120 grams.
Cocoanut oil, or sweet almond oil	120 grams.

Rub in with a soft sponge once or twice daily.

General Lotions for Baldness.

It must not be forgotten that the causes of baldness are many, and it by no means follows that a preparation which works wonders in certain cases is equally effective in others. Of course, special examination and treatment are to be preferred to the application of general preparations. However, there is a demand for the latter. If the baldness is brought about by some severe illness, such as typhoid fever, there is a good chance of the roots of the hair being stimulated to renewed growth. If it is due to seborrhœa, the treatment may commence with a scurf lotion such as is indicated above. If due to the effects of some chronic disease, there is no remedy unless the disease itself can be at least temporarily cured. For such cases one of the following may be tried:

I.—Borax	1 dram.
Glycerin	2 drams.
Tincture of cantharides	1 oz.
Potassium carbonate	½ dram.
Bay rum	1 oz.
Distilled water	to 10 oz.
II.—Pilocarpin hydrochloride	10 grains.
Rose water	3 oz.
Eau de cologne	1 oz.
III.—Lactic acid	3 drams.
Castor oil	2 drams.
Lavender water	4 drams.
Alcohol	to 4 oz.

Dandruff Removers.

Probably the most effective application for dandruff is sulphurous acid (not *sulphuric*). The hair should be first well cleansed with a shampoo liquid and the lotion well rubbed into the scalp. It may be used daily until the scurf is thoroughly removed, after which an application once a week should be effective in preventing further development. The lotion is prepared as follows:

Sulphurous acid.................. .... 2 oz.
Salicylic acid.............................. 30 grains.
Alcohol................................... 4 oz.
Water.................................to 10 oz.

Dissolve the salicylic acid in the alcohol, add the sulphurous acid, and finally the water. This is quite harmless to the hair and the skin, but should not be used after the application of any dye.

B. Hair Restorers.

These preparations, which are claimed to restore the color of the hair, are actually dyes generally containing lead and sulphur, which combine to deposit sulphide of lead on the hair. They not only have no beneficial effect in restoring the natural color, but are generally considered dangerous to use on account of their liability to cause lead poisoning. Most proprietary hair restorers are included in this class.

The following is an imitation of a popular hair restorer:

Lead acetate............................ ¾ oz.
Precipitated sulphur...................... 1½ oz.
Glycerin................................. 5 oz.
Essence of heliotrope..................... 1 oz.
Distilled water to produce................. 1 pint.

Dissolve the lead acetate in half of the water, rub the sulphur with the glycerin in a mortar and add the lead solution gradually. Then add the perfume and water to the required volume. This preparation must be labeled, "Shake the bottle."

Another formula is as follows:

Acetate of lead........................... 2 drams.
Sodium hyposulphite...................... 4 drams.
Glycerin................................. 2 oz.
Alcohol.................................. 1 oz.
Rose water to produce.................... 1 pint.

Dissolve the lead acetate in 2 ounces and the hyposulphite in 4 ounces of the rose water. Add the latter to the former, then add the other ingredients in the order named and filter.

SAGE AND SULPHUR HAIR RESTORER.

This is an old-fashioned remedy which has been in favor for many years and for which there is a large popular demand. It was formerly prepared in a very crude manner by adding precipitated sulphur to an infusion of sage leaves, but in modern practice milk of sulphur, which is more finely divided than the ordinary form, is used and several other ingredients are added, including acetate of lead, which combines with the sulphur to darken the hair, as already noted. The acetate of lead may be omitted if desired. The complete formula is as follows:

Sage leaves..............................	1 oz.
Henna leaves............................	½ oz.
Milk of sulphur.........................	3 oz.
Acetate of lead.........................	1½ oz.
Tincture of cantharides.................	2 oz.
Glycerin................................	1 pint.
Boiling water...........................	1 gallon.

Pour the boiling water over the sage and henna leaves and let stand until cool, then strain. Rub the sulphur and the acetate of lead together and add the glycerin and cantharides slowly to make a paste. Then stir into the sage and henna liquid. Color and perfume if desired. As sulphur is insoluble, this will make a cloudy mixture which must be labeled "To be shaken before using."

CHAPTER XXXII.

POMADES AND HAIR OILS.

THE hair, the beautiful ornament of the human body, requires fat for its care and preservation, for there are but few
persons whose scalp is so vigorous that the hair can derive
sufficient nourishment from it to maintain its gloss and smoothness.

Among the ancient Greeks, Romans, and Germans various
ointments were in use for the care of the hair. In Rome
there was even, as we have stated in an earlier part of the
book, a special guild of ointment-makers or unguentarii.
They employed a process for making their ointments fragrant
which resembles that of maceration in present use.

The so-called pomades (from pomum, apple) were prepared
by sticking a fine apple full of spices and placing it for a long
time in liquid fat which absorbed the odor of the spices.

In the present state of chemical science, the basis of every
pomade or hair oil is formed by some fat perfumed with aromatic substances and at times colored. The fats generally
used are lard, beef marrow, tallow, bears' grease, olive or almond oil; some of the firmer fats receive an addition of a certain amount of paraffin, spermaceti, or wax, in order to give
the pomade greater consistence. As in the manufacture of

all the finer articles, it is essential that whatever fat is employed should be perfectly pure; only fat which is absolutely neutral, *i.e.*, free from acid, can be used, and any sample with but a trace of rancidity (containing free fatty acids) should be rejected on account of the penetrating odor peculiar to several of these acids.

Manufacturers who aim at the production of fine goods spare neither trouble nor expense in order to obtain perfectly pure fats.

Fats are purified for the purposes of the perfumer in the following manner:

The fat is melted in a bright iron pot or enamelled vessel with three times the quantity of water containing in solution about one per cent (of the weight of the fat) of alum and one per cent of table salt. Fat and water are well stirred with a broad flat ladle or some mechanical arrangement within the boiler. After the mass has remained at rest for some time, the curdled solid matters are skimmed from the surface. The time required for this operation can be much shortened by the use of a pump which raises the fat and water from the boiler and returns them in a fine spray.

When fats with some degree of rancidity are to be made suitable for the purposes of the perfumer, 0.5% of caustic soda lye is added to the water instead of the alum.

After this treatment is completed, the fat must be washed in order to free it from the substances with which it was purified. Formerly this washing was done in a manner resembling the grinding of oil colors. The fat was placed on a level stone plate and kneaded with a muller with flat base under a continual stream of water flowing from above, until the fat was clean. This expensive hand labor is now performed by machines, the fat being treated with water in vertical mills.

No matter how carefully a fat was purified, it may happen that the pomades made from it, if kept long in stock, may subsequently become rancid—a circumstance which may destroy the reputation of a factory. Fortunately we know two sub-stances which materially counteract the tendency of fats to become rancid: salicylic acid and benzoin. Either of these substances is added to many perfumery articles, especially pomades, in order to prevent rancidity; an admixture of from one-one-thousandth to five-one-thousandths parts of solid salicylic acid suffices, according to our experiments, for the purpose; of benzoin we need about three-fourths of an ounce for every quart of fat; the resin is only partly soluble in fat, but imparts to it its vanilla-like odor. For the finest pomades sublimed benzoic acid is used, in the proportion of about 150 to 240 grains to the quart of fat.

CHAPTER XXXIII.

FORMULAS FOR THE MANUFACTURE OF POMADES AND HAIR OILS.

A. Pomades.

IN manufacturing perfumery two groups of pomades are distinguished—those with a hard base, and those with a soft base. By base is meant the fat which is the vehicle of the odor in every pomade. The consistence of the substance de-pends upon its melting-point; lard and beef marrow, having a low melting-point, furnish soft pomades; while beef and mut-ton tallow, which often receive an addition of paraffin, wax, or spermaceti in order to make them firmer, have a higher melting-point and serve for hard pomades.

French perfumers put on the market some very fine po-mades consisting of the fat which has served for the absorption of odors by maceration, enfleurage, etc., and which has been treated with alcohol for the extraction of the odors (so-called washed pomades). No matter how long such a fat is treated with alcohol, it tenaciously retains a portion of the odor to which the great fragrance of these pomades is due and which has given them their reputation.

If the pomades resulting from the following formulas should turn out too soft—a fact depending on the climate of the place of manufacture—they may receive an addition of a mixture of equal parts of paraffin, wax, and spermaceti, in portions of respectively five per cent at each addition, until the desired ointment-like consistence is attained.

GENERAL HAIR POMADES.

These are used for lubricating the scalp where there is a deficiency of natural fats. The principal ingredients of these general hair pomades are the following: Petrolatum, lard, palm oil, almond oil, peach-kernel oil, castor oil, cocoanut oil, lanolin, beeswax, and spermaceti, the last two being added to stiffen the pomade.

All animal and nearly all vegetable oils have a tendency to become rancid, and if required to be kept for any length of time a preservative should be added. Benzoic acid is the most suitable preservative, which may be added in the proportion of about 10 grains to the pound, but the same effect is produced by steeping half an ounce of powdered benzoin in a pound of the oil or fat for an hour in a bath of boiling water, stirring frequently. Oils and fats treated in this way are termed "benzoated." The following are typical recipes for hair pomades:

Olive oil	8 oz.
Benzoated lard	7 oz.
White wax	1 oz.
Oil of rose geranium	30 minims.
Oil of bergamot	2 drams.
Oil of neroli	20 minims.

Melt the first three ingredients together, stir till the mass begins to set, then add the essential oils.

TRANSPARENT POMADE.

Spermaceti............................... 2 oz.
Castor oil................................. 4 oz.
Alcohol................................... 4 oz.
Oil of bergamot.......................... 1 dram.
Oil of neroli............................. 30 minims.
Oil of cloves............................. 10 minims.

Melt the spermaceti, add the castor oil, and then the alcohol in which the essential oils have been previously dissolved. Fill into pots and allow to cool without stirring.

CRYSTALLINE POMADE.

Spermaceti............................... 1 oz.
Petrolatum............................... 4 oz.
Castor oil................................ 4 oz.
Oil of cassia............................ 30 minims.
Oil of cloves............................. 10 minims.
Oil of bergamot.......................... 1 dram.

Melt the spermaceti, add the oils and allow to cool without stirring.

LANOLIN POMADE.

Anhydrous lanolin........................ 1 oz.
Petrolatum............................... 3 oz.
Tincture of benzoin...................... 1 dram.
Heliotropin.............................. 10 grains.

Petrolatum does not become rancid and hence is generally preferred for hair preparations, though it is not so readily absorbed by the scalp as are the vegetable oils. Sometimes a

mixture of mineral and vegetable fats is employed as a basis
or hair pomades. An example of this type is the following:

Benzoated lard	4 oz.
White wax	1 oz.
Petrolatum	4 oz.
Oil of bergamot	1 dram.
Oil of lavender	30 minims.
Otto of rose	10 minims.

MEDICATED POMADES.

Medicated pomades are used for dandruff, for baldness, and
or the destruction of parasites.

DANDRUFF POMADE.

Salicylic acid	2 drams.
Boric acid	1 dram.
Vaseline	4 oz.
Oil of cinnamon	10 minims.
Oil of bergamot	20 minims.

Rub the salicylic acid and the borax in a mortar with a
little of the vaseline, then add the remainder of the vaseline and
the essential oils.

POMADES FOR BALDNESS.

These pomades usually contain either pilocarpin salts or
cantharides. Quinin salts are sometimes used, but are of
doubtful value. The following is a formula for a stimulating
pomade:

Pilocarpin hydrochlorate	10 grains.
Balsam of Peru	30 grains.
Precipitated sulphur	1 dram.
Benzoated lard	1 oz.

Dissolve the pilocarpin in a few drops of water and mix in a mortar with the other ingredients.

STAVEACRE POMADE.

This is very effective in destroying vermin, and accordingly a formula for it is appended:

Staveacre seeds, crushed..................	4 oz.
Benzoated lard...........................	1 lb.
White wax...............................	4 oz.

Melt the lard over a water bath, add the staveacre seeds and digest for two hours. Strain through cloth, pressing strongly, and add the wax previously melted. Stir continuously until cold.

ANTI-KINK HAIR POMADE.

Beef suet..................................	16 oz., av.
Yellow wax...............................	2 oz., av.
Castor oil.................................	2 oz., av.
Benzoic acid.............................	10 grains.
Oil of lemon.............................	1 fl. dram.
Oil of cassia..............................	15 drops.

Mix the suet and wax, add the castor oil and acid, allow the mixture to cool somewhat, and incorporate the other oils. This is used to straighten kinky hair and make it lie flat.

POMADE FOR BALDNESS.

Pilocarpin hydrochlorate...................	2 parts.
Quinin hydrochlorate......................	4 parts.
Precipitated sulphur......................	10 parts.
Balsam of Peru...........................	20 parts.
Beef marrow, to make.....................	100 parts.

Mix. Apply to the head after the use of a saponaceous lotion.

HAIR OILS.

These are not as much used now as formerly, but there still is a demand for them. They usually consist of benzoated oil to which any desired floral odor has been added. Benzoated oil is made by digesting an ounce of bruised benzoin, Siam preferably, in a pint of almond or olive oil for three hours on a water bath, and filtering through French gray paper. Oil so prepared does not become rancid.

HELIOTROPE.

Benzoated oil............................	30 oz.
Heliotropin...............................	½ dram.

ROSE.

Benzoated oil............................	20 oz.
Otto of rose..............................	20 minims.
Oil of rose-geranium......................	10 minims.

VIOLET.

Benzoated oil............................	10 oz.
Ionone, 100%.............................	2½ oz.
Otto of rose..............................	2 minims.
Oil of jasmine, syn.......................	3 minims.
Oil of cloves.............................	6 minims.
Oil of bergamot...........................	12 minims.

Mix. For other odors mix 1 part of any floral oil with 4 parts of benzoated oil.

BAY RUM.

While the well-known bay rum is used more as a face lotion or refreshing skin tonic, particularly after shaving or when perspiring in hot weather, yet it is also often used as a wash for the scalp, and is popularly believed to stimulate the growth of hair, which is in reality not the case. We shall therefore give some formulas for its preparation here:

Bay Rum.

Oil of bay (from Myrcia acris)............	240 grains.
Oil of orange (bigarade).................	16 grains.
Oil of pimenta.........................	16 grains.
Alcohol...............................	1 qt.
Water................................	25 fl. oz.

Dissolve the oils in the alcohol and add the water. Mix the liquid with about 2 ounces of precipitated phosphate of lime, and filter. It will improve with age.

A cheaper preparation suitable for barbers' use may be made as follows:

Bay Rum for Barbers' Use.

Oil of bay.............................	3 drains.
Alcohol...............................	2 pints.
Tincture capsicum......................	2 drams.
Water................................	6 pints.

Color to suit. Allow to stand twenty-four hours and then filter. After that add 1 dram acetic ether.

Ammoniacal Bay Rum.

Bay rum..............................	10 oz.
Tincture of cantharides.................	2 oz.
Strong solution of ammonia..............	2 drams.
Glycerin..............................	1 dram.
Water............................to	1 pint.

The ammonia assists in removing grease, with which it forms a soap.

Genuine bay rum is imported from the West Indies (St. Thomas, etc.), where a crude kind of alcohol, obtained in connection with the manufacture of rum from molasses, is distilled.

with the fresh leaves of the bay-tree (Myrcia acris). The oil
of bay obtained from this must not be confounded with the
oil of sweet bay. The latter, as it appears in commerce, is a
crude mixture of a fixed with a volatile oil.

CHAPTER XXXIV.

HAIR DYES AND DEPILATORIES.

THE custom of dyeing the hair is universal in the Orient;
in the Occident, however, hair dyes are also frequently used,
namely, to hide the grayness of the hair, sometimes to give
the hair a preferred color. Hair dyes, which are very numerous, may be divided into groups—those containing the dyestuff ready formed, and those in which it is produced in the
hair by some chemical process. Some hair dyes contain substances which in their nature are decidedly injurious to the
hair; such articles, of course, must be dispensed with because,
if frequently employed, they would certainly lead to baldness. We shall return to this subject in connection with the
several preparations.

Regarding the use of hair dyes, especially those consisting
of two separate portions, we may state that it is necessary to
remove the fat from the hair before applying the dye, as the
chemicals in question do not adhere well to fat. The hair
should be thoroughly washed once or twice with soap, and
dyed when nearly dry.

When dyeing the hair the preparations should first be diluted; if the color is not deep enough, the process is repeated.
If the preparation is used at once in a concentrated form, a
color may result which has no resemblance to any natural
tint; hair meant to be black may assume a metallic bluish-
black gloss.

A. Simple Hair Dyes.

WALNUT HAIR DYE.

One of the oldest forms of hair dye or stain is the juice of green walnut shells, which, as everyone knows who has ever gathered walnuts, produces a rich dark-brown stain upon the skin. This stain is very difficult to remove from the skin, but is not so easily applied to the hair, as it is necessary to remove the oil from the hair by shampooing, and also to use alum or some similar ingredient with the walnut juice in order to fix the color. When these precautions are observed the stain is said to produce very satisfactory results.

HAIR DYE FROM WALNUT SHELLS.

Green walnut shells.......................450 grams.
Powdered alum........................... 30 grams.
Rose water...............................120 grams.

The ingredients are triturated together in a mortar, pressed, and treated with 90 per cent alcohol in the proportion of 30 parts of alcohol to 100 of liquid. The mixture is then left for four days in a close vessel and finally filtered, and the liquid perfumed to suit.

WALNUT EXTRACT.

If it is desired to make an extract from walnut shells for subsequent use, they are pounded with a pestle and then covered with water containing 1 per cent of salt. After three days the whole is poured into a large pan, on which a mark is then made to show the height of the liquid, it being necessary to replace the water lost by evaporation. Heat to near boiling-point for four to six hours, allow it to cool, and press out the liquid. In the absence of a press, this may be done with the aid of a linen cloth, or preferably a sack of canvas, about 40 inches long and

20 inches wide, which is half filled with the mass from the pan, held over an open vessel, tied up at the mouth, and twisted by means of a couple of sticks, two persons being required for the operation. Nevertheless, the bag must not be twisted so tightly as to cause it to burst. The liquid thus obtained is returned to the pan, and is concentrated to one-fourth its initial volume, which is measured for that purpose, the best plan being to place a quantity of water equal to one-quarter of the liquid in the pan, marking the level, pouring out the water again, and then allowing the nut liquor to evaporate until the level of the mark is reached. The finished extract then receives an addition of 16 per cent of 95 per cent alcohol, and is either stored in tightly closed vessels, for stock, or finished off, ready for use, by the addition of perfume as desired.

PYROGALLIC HAIR DYE.

An innocuous dye, to be applied with the brush morning and evening after using an alkaline wash, is composed of

Pyrogallic acid............................	4 grams.
Citric acid................................	3 grams.
Boroglycerin..............................	10 grams.
Distilled water............................	90 grams.

IRON-SULPHATE HAIR DYE.

Iron sulphate.............................	2 drams.
Glycerin..................................	1 ounce.
Eau de cologne............................	1 ounce.
Rose water...............................	14 ounces.

AUREOL HAIR DYE.

Aureol for dyeing the hair is said to be harmless, is very stable, and the color is not altered, even by soap. It consists of 1 per cent of methol, 3 per cent of amidophenol chlorate, and 6 per cent of monoamidphenylamine, all dissolved in 50 per cent of alcohol containing 5 per cent of sodium sulphite.

The hair must be first washed with soap and water, to remove all grease. The aureol is then mixed with an equal quantity of hydrogen peroxide, and applied, by means of a fine comb, to the still damp hair. At the end of two or three hours the hair assumes a handsome dark-brown color, which can be made still darker by repeating the application as desired.

Kohol (Teinture Chinoise).

Gum arabic.............................	1 ounce.
India ink...............................	1¾ ounces.
Rose water.............................	1 quart.

Powder the ink and the gum, and triturate small quantities of the powder with rose water until a uniform black liquid results, which must be free from granules. This liquid is placed in a bottle and the rest of the rose water added. Kohol can be used only by persons with black hair, and is employed particularly for dyeing the eyebrows. As the coloring matter of this preparation consists of carbon in a state of fine division, the dye is perfectly harmless, but it is removed by washing.

B. Two-Solution Dyes.

Silver Hair Dyes.

These consist of two preparations contained in bottles marked I. and II. respectively. The first contains a solution of sulphide of potassium, pyrogallic acid, sulphide of sodium, etc., in alcohol. The second bottle contains the silver solution, and should be of dark amber-colored or black glass as the silver salts are decomposed by light. Blue glass can not be employed for this purpose. In use, some of the liquid from bottle I. is poured into a cup, and the hair is moistened with it by means of a soft brush.

The liquid from bottle II. is poured into a second cup and applied with another brush.

SILVER DYE I.

I. (*In White Bottle.*)

Sulphide of potassium...................... 7 oz.
Alcohol................................... 1 qt.

II. (*In Dark Bottle.*)

Silver nitrate............................. 4¼ oz.
Distilled water........................... 1 qt.

The sulphide of potassium (liver of sulphur) appears in fragments of a liver-brown mass which readily dissolves in water. The solution must be filtered before being filled into bottles for sale, and, as it becomes turbid in the air, kept in well-closed vessels. When the two solutions are brought together, black sulphide of silver results and darkens the hair. After the use of this preparation a disagreeable odor of rotten eggs adheres to the hair, but can be easily removed by washing, especially with one of the previously mentioned hair washes.

The silver hair dye will be still better if the liquid contained in bottle II. is made by dropping into the solution, under continual stirring, ammonia water, until the precipitate first formed is again dissolved, as in the following:

SILVER DYE II.

I. (*In White Bottle.*)

Pyrogallic acid........................... 15 grains.
Alcohol of 40%........................... 1 pint.

II. (*In Dark Bottle.*)

Silver nitrate............................150 grains.
Distilled water........................... 2¾ oz.
Ammonia water........................... 1 oz.

SILVER DYE III.

I. (*In White Bottle.*)

Sulphide of sodium........................120 grains.
Distilled water............................ 3½ oz.

II. (*In Dark Bottle.*)

Silver nitrate.............................. 45 grains.
Distilled water............................ 3½ oz.

TANNIN AND SILVER HAIR DYE.

I. (*In White Bottle.*)

Powdered nut-galls........................ 14 oz.
Water..................................... 1 pint.
Rose water................................ 1 pint.

Boil the nut-galls in the water, strain the boiling liquid through a thick cloth into the rose water, and fill the still hot mixture into bottles which must be immediately closed. (It is essential that the liquid be hot during the filling, to guard against the development of mould.)

II. (*In Dark Bottle.*)

Silver nitrate.............................. 5½ oz.
Distilled water............................ 1 qt.

Add ammonia water to the silver solution until the precipitate first formed is again dissolved.

COPPER HAIR DYE.

I. (*In White Bottle.*)

Ferrocyanide of potassium................. 7 oz.
Distilled water............................ 1 qt.

II. (*In Dark Bottle.*)

Sulphate of copper........................ 7 oz.
Distilled water............................ 1 qt.

Add ammonia water to the copper solution until the light blue precipitate first formed again dissolves to a rich dark-blue liquid. This hair dye gives a dark-brown color.

EAU DE FONTAINE DE JOUVENCE,

also called Auricome and Golden Hair Water, is no dye, but a bleaching agent which changes dark hair to a light blond or golden-yellow color. The preparation consists of peroxide of hydrogen, a substance possessing marked bleaching properties.

Peroxide of hydrogen, or hydrogen dioxide, is at the present time made on a large scale by many manufacturers, and readily obtainable in the market. It would therefore scarcely pay any one to prepare it himself unless he were out of reach of the usual channels of trade, so that he could not obtain the preparation in a *fresh* state. Nevertheless it may be useful to state how it is made. Barium dioxide (or peroxide), which is a regular article of commerce, and is a stable compound which will keep for any length of time if kept in tightly closed bottles, is treated with water until the dioxide forms with it a thin, smooth milk. This is gradually added to dilute sulphuric acid, cooled with ice or kept otherwise as cold as possible, until the sulphuric acid is almost entirely neutralized. The solution is then allowed to settle and the clear liquid drawn off. For bleaching purposes, this is pure enough. Only it must be ascertained that the amount of free acid present, without which the hydrogen dioxide does not keep well, is only small. Other acids can be used besides sulphuric, but the latter is the most convenient. If an alkali is added to hydrogen dioxide so that the reaction becomes alkaline, it will decompose very rapidly. Even under the most favorable circumstances (when acid, and kept in a cool place) it will gradually deteriorate, and finally be entirely converted into oxygen gas, which escapes, and plain water.

Peroxide or dioxide of hydrogen, when applied to the hair as a bleaching agent, must be used in a dilute condition at first. Those who use it for the first time should always make preliminary trials with the liquid upon odd bunches of hair (such as may at any time be procured at hair-dressers' shops) resembling that which is to be bleached, before actually applying it to the latter.

The hair to be bleached is deprived of fat by washing with soap solution, the soap is washed out with water, and the peroxide of hydrogen applied.

WHISKER DYE.

I. Acetate of lead	1¾ oz.
Distilled water.................... ...	1 pint.
II. Caustic potassa........	¾ oz.
Distilled water....................	1 qt.

Dissolve the acetate of lead ("sugar of lead") in the warm water, filter the solution, and add ammonia water until a precipitate ceases to form.   Collect the precipitate on a filter, wash it by pouring distilled water over it eight or ten times, and while still moist introduce it into solution II.   Stir repeatedly, and after twelve hours leave the vessel at rest until the solution has become clear.   Then decant it from the sediment, which may be treated a second time with solution II. For use, the beard is washed with soap, and combed with a fine rubber comb dipped in the solution.

C. Depilatories.

Combinations of sulphur with the alkaline metals calcium, barium, and strontium rapidly destroy the hair; for this reason tanners use the "gas lime" from gas works, which contains calcium sulphide, for removing the hair from hides.   All the depilatories used cosmetically, even rhusma employed in the Orient for removing the beard, owe their activity to the presence of calcium sulphide.

CALCIUM SULPHIDE

has usually been lauded as a perfectly harmless depilatory. This is a great mistake, however, since it has often done serious harm, through careless application by persons unfamiliar with its caustic and corrosive effects. It is absolutely necessary to protect the *skin* against its action; otherwise superficial irritation, or even destruction of the skin may result.

Calcium sulphide cannot be made by the action of sulphuretted hydrogen upon lime. It is usually made by heating at a low red heat, in a securely closed crucible, an intimate mixture of 100 parts of finely powdered quicklime with 90 parts of precipitated sulphur. Mix together:

Calcium sulphide........	4 oz.
Sugar........	2 oz.
Water.....................	2 oz.
Starch powder..............	2 oz.
Oil of lemon.........	30 grains.
Oil of peppermint........................	10 grains.

The resulting mass must be filled at once into an air-tight jar, as the calcium sulphide is decomposed in the atmosphere. For use, some of the mass is moistened with water, painted on the skin, and washed off with water after thirty to forty-five minutes. This and all other depilatories act only temporarily, that is, they destroy only the hair projecting above the surface without killing the hair bulbs; after some time the hair grows again and the preparation must be reapplied.

BARIUM SULPHIDE,

which is likewise used as a depilatory, is made by heating barium sulphate with charcoal, extracting the residue with water, and mixing the resulting product with starch paste. In its effects barium sulphide equals the preceding preparation, but it decomposes more readily.

DEPILATORY PASTE.

Powdered caustic lime..................	2 lb.
Starch powder........................	2 lb.
Sodium sulphide......	21 oz.

Sodium sulphide is made by saturating strong caustic soda solution with sulphuretted hydrogen. The other ingredients are added to the solution of sodium sulphide.

DEPILATORY POWDER.

Barium sulphide............................. 1 oz.
Zinc oxide................................... 1 oz.
Powdered starch .... 1 oz.

Make into paste with a little water when required for use.

RHUSMA.

is a depilatory made by mixing powdered quicklime (unslaked) with orpiment (yellow sulphide of arsenic), and used in the Orient. Owing to its poisonousness and the destructive effects of the caustic lime on the skin, this preparation should never be employed.

CHAPTER XXXV.

WAX POMADES, BANDOLINES, AND BRIL-LIANTINES.

THE so-called wax pomades, stick pomatum, and bandolines serve to stiffen the hair and are frequently employed by hair dressers. The former two articles possess some adhesive power by which they fasten the hair together; bandolines are mucilaginous fluids which generally contain bassorin (or vegetable mucilage present in tragacanth), quince seeds, etc.

A. Wax Pomades.

STICK POMATUM.

This is usually formed into oval or round sticks which are wrapped in tin foil. They are colored and perfumed as desired. The ordinary varieties are: white, for light blond hair, which is left uncolored; pink, colored with carmine; brown, colored with umber; and black, colored with bone black. The coloring matters are always rubbed up with oil. Red pomatum may be colored with alkanet root, which is macerated for some time with the melted fat. The base of these preparations consists of:

Lard.................................. ...   4 lb.
Tallow.................................   12 lb.
Wax....................................   6 lb.

The mass may be made harder or softer by increasing or diminishing the wax. The perfumes generally used are oils of bergamot, lemon, clove, and thyme, with an addition of some Peru balsam.

B. Beard Wax.

BEARD WAX (CIRE À MOUSTACHES).

Turpentine...........................   2 lb.
Expressed oil of almond..............   2 lb.
Wax..................................   6 lb.
Violet pomade........................   2 lb.
Peru balsam..........................   1 lb.
Oil of clove.........................   1 oz.
Oil of santal.... ....   ¾ oz.
Oil of cinnamon......................   ¾ oz.

HUNGARIAN BEARD WAX (CIRE À MOUSTACHE HONGROISE.)

Castile soap, powdered...............   3½ oz.
Mucilage of acacia...................   10 oz.
White wax............................   9 oz.
Glycerin.............................   3½ oz.
Oil of bergamot......................   20 drops.
Oil of lemon.........................   10 drops.
Oil of rose..........................   10 drops.

Rub the powdered soap with the mucilage, previously diluted with nine ounces of water, then add the wax and glycerin, and heat the mass on a water-bath, stirring constantly, until it becomes homogeneous. Lastly add the oils, and pour the mass into suitable moulds.

For brown or black wax the corresponding color is added. The mass is formed into sticks the thickness of a lead pencil.

C. Bandolines.

BANDOLINE AUX AMANDES.

Tragacanth............................. 14 oz.
Rose water............................. 8 qts.
Oil of bitter almond.................... ¾ oz.

Crush the tragacanth, place it in the rose water, and leave it at rest in a warm spot, stirring occasionally, until the traga- canth has swollen to a slimy mass. Press it first through a coarse and then through a finer cloth, add a little carmine and the oil of bitter almond.

BANDOLINE À LA ROSE.

This is made like the preceding, only substituting 1½ oz of oil of rose for the oil of bitter almond. Other varieties may be produced by the use of different odors.

D. Brillantines.

Under various names preparations are placed on the mar- ket which render the hair both soft and glossy. The chief constituent of all these articles is gylcerin which is perfumed according to taste and stained reddish or violet. As many aniline colors easily dissolve in glycerin, they are generally used for this purpose. Formerly, before glycerin was obtain- able in sufficient purity, brillantines were chiefly made of cas- tor oil dissolved in alcohol, but on account of the fact that glycerin is cheaper than castor oil with alcohol, the former is preferable.

Brillantine.

Glycerin.... .... 8 lb.
Extract of jasmine (or other flower)........ 2 qts.

Oléolisse.

Glycerin.......................... ... 4 lb.
Castor oil................... .... 4 lb.
Oil of bergamot....................... .... ¾ oz.
Oil of lemon......... .... ¾ oz.
Oil of neroli....................... ....150 grains.

CHAPTER XXXVI.

SKIN COSMETICS AND FACE LOTIONS.

THE use of skin cosmetics and paints is of remote anti-quity, but varies in different nations according to their civi-lization and their sense of beauty.   While among certain Oriental nations dark blue rings around the eyes, with yellow lips and nails, pass for beautiful, the European prizes only a white skin with a delicate tinge of red; Italian ladies in the middle ages used the dark red juice of the fruit of the deadly night-shade as a paint, hence the name bella donna, *i.e.*, beau-tiful lady. (According to Matthiolus, the name *herba bella donna* arose from the fact that Italian ladies used a distilled water of the plant as a cosmetic.)   Owing to its marked effect on the eyes, by dilating the pupil and increasing the lustre, this juice also heightens the brilliancy of the eye, though at the expense of its health.

While in the last century face-painting was a universal fashion, it is nowadays resorted to only by persons whose skin requires some artificial help.   But nobody desires that the cosmetic should be perceptible on the skin.   Hence it must be laid down as a rule that paints and all cosmetics should be so compounded that it is not easily possible to the observer to recognize that some artificial means has been employed for beautifying the skin.

We give below a number of such articles, which come as near as possible to this ideal without injuring the skin.   As every skin cosmetic cannot but occlude the pores of the skin, it should be removed as soon as possible—an advice to be heeded particularly by actors and actresses, who must appear painted on the boards.

A. White Skin Cosmetics.

FRENCH WHITE (BLANC FRANÇAIS).

```
Talcum.................................  4 lb.
Oil of lemon...........................  75 grains.
Oil of bergamot........................  75 grains.
```

The talcum must be reduced to the finest powder, levigated, dried, and then perfumed. Owing to its unctuous nature, it readily adheres to the skin, and as it has no effect on it and does not change color, it is the best of all powders.

LIQUID BISMUTH WHITE; PEARL WHITE (BLANC PERLÉ LIQUIDE).

```
Subnitrate of bismuth..................  1 lb.
Rose water.............................  1 qt.
Orange-flower water....................  1 qt.
```

When standing at rest, the subnitrate of bismuth sinks to the bottom, while the supernatant fluid becomes quite clear. The bottle must therefore be vigorously shaken immediately before use. When this preparation remains on the skin for some length of time, it loses its pure white color and becomes yellow, or darker, through the gradual formation of a black sulphur compound.

VENETIAN CHALK (CRAIE VENÉTIENNE).

is made exactly like the French white, above; the only difference between the two preparations is that the talcum for the latter is brought to a red heat, which, however, causes it in part to lose the power of adhering to the skin.

B. Red Skin Cosmetics (Rouges).

Rouge Végetal Rose Liquide.

Ammonia water......................	2 oz.
Carmine...........................	1¼ oz.
Essence of rose (triple)...............	2½ oz.
Rose water.........................	2 qts.

This superior preparation, which serves mainly for coloring the lips, is made as follows: Reduce the carmine to powder; macerate it in the ammonia in a three or four pint bottle for several days, add the other ingredients, and let it stand
for a week under oft-repeated agitation. At the end of that
time the bottle is left undisturbed until the contents have become quite clear, when they are carefully decanted and filled
into bottles for sale.

In order to obtain this preparation in proper form, only
the finest carmine should be used. That known in the market as " No. 40 " is the best. This alone will produce a cosmetic that, when brought in contact with the skin, will give
a vivid red color.

In place of carmine, which requires the presence of ammonia if it is to remain in solution, the anilin color known as
eosine may be used. Of this, very minute amounts will be sufficient to impart the proper tint. It is impracticable to give
exact proportions, as these must be determined in each case
by experiment. It is necessary to avoid an excess. The tint
of a liquid colored by eosine may not appear deep, and yet
when it is applied to the skin a decidedly deeper stain than
was desired may be produced. Hence each addition of fresh
coloring matter must be carefully controlled by a practical test.

Rouge en Feuilles.

Cut from thick, highly calendered paper circular disks
about 2½ inches in diameter, and cover them with a layer of

carmine containing just enough gum acacia to make it adhere to the paper. For use, the leaf is breathed on, a pledget of fine cotton is rubbed over it, and the adhering color is transferred to the skin.

ROUGE EN PÂTE.

Carmine............................	1 oz.
Talcum.............................	21 oz.
Gum acacia.........................	1¾ oz.

The ingredients in finest powder are mixed in a mortar by prolonged trituration, then water is added in small portions to form a doughy mass to be filled into shallow porcelain dishes about the diameter of a dollar. If the rouge is desired darker for the use of actors and dark-complexioned persons, the proportion of carmine should be increased.

ROUGE EN TASSES.

Carthamin	1 oz.
Talcum powder......................	1 lb.
Gum acacia.........................	1½ oz.
Oil of rose........................	15 grains.

This rouge, when dry, has a greenish metallic lustre; it is prepared and sold like rouge en pâte.

BLEU VÉGÉTAL POUR LES VEINES.

Venetian chalk.....................	1 lb.
Berlin blue........................	1¾ oz.
Gum acacia	1 oz.

To the powdered solids add sufficient water to form a mass to be rolled into sticks. For use, a pencil is breathed on, rubbed against the rough side of a piece of white glove leather, and the veins are marked with the adhering color on the skin coated with pearl white. Of course, some dexterity is required to make the veins appear natural by the use of this blue color.

18

Rouge Alloxane (Alloxan Red; Murexide Paint).

Cold cream.......................... 1 lb.
Alloxan............................. 75 grains.

Dissolve the alloxan in a little water and mix it intimately with any desired cold-cream. The mixture is white, but when transferred to the skin gradually becomes red. The preparation sold in Austria, etc., under the name of "Schnuda" is identical with this alloxan paint.

C. Face Lotions.

The skin often contains spots with marked color which are more or less unsightly; for instance, freckles, liver spots, mother's marks (nævi), etc. Unfortunately we know of no remedy which radically removes them; even chemical preparations with the most energetic effects, which of course must never be employed owing to their destructive action on the skin, cannot entirely do away with these dark spots which have their seat in the lower layers of the skin. But the public demands preparations for the removal of freckles, liver spots, etc., and—obtains them. We subjoin the formulas for several of such secret remedies, but declare emphatically that none of them will completely effect the desired result.

Freckle Milk (Lait Antéphelique).

Camphor .... 1¾ oz.
Ammonium chloride.................. ¾ oz.
Corrosive sublimate....................150 grains.
Albumen............................. 3½ oz.
Rose water.......................... 2 lb.

We call attention to the fact that the sublimate (bichloride of mercury) is very poisonous and must be used with the greatest care.

FRECKLE LOTION.

Angelica root	1¾ oz.
Black hellebore root	1¾ oz.
Storax	¾ oz.
Oil of bergamot	150 grains.
Oil of citron	150 grains.
Alcohol	2 qts.

Macerate for a week and filter.

EAU LENTICULEUSE.

Potassium carbonate	7 oz.
Sugar	¾ oz.
Orange-flower water	2 qts.
Alcohol	7 oz.

LILIONESE I.

Potassium carbonate	14 oz.
Water	4 lb.
Rose water	14 oz.
Alcohol	7 oz.
Oil of rose	150 grains.
Oil of cinnamon	75 grains.

LILIONESE II.

Rose water	2 qts.
Orange-flower water	1 qt.
Glycerin	1 lb.
Potassium carbonate	3½ oz.
Tincture of benzoin	¾ to 1¾ oz.

Add only enough of the alcoholic tincture of benzoin to render the liquid slightly opalescent or milky.

LOTION FOR CHAPPED SKIN.

Glycerin	4 lb.
Water	1 qt.
Rose water	1 qt.

Color pale red with cochineal.

EAU DE PERLES.

White soap............................ 1 lb.
Dissolved in: Water.................... 4 qts.
          Glycerin.................. 2 lb.
Add: Rose water...................... 1 qt.
         Tincture of musk................150 grains.

To be colored bluish with some indigo-carmin.

TEINT DE VENUS.

Alcoholic soap solution.................. 2 qts.
Carbonate of potassium.................. 3½ oz.
Extract of orange flower............... 3½ oz.

The soap solution is made as concentrated as possible, and the entire fluid colored with cochineal; in place of the extract of orange flower, other essences or extracts may also be employed. For use, some of the liqiud is poured into the wash water.

PULCHÉRINE.

Carbonate of potassium................ 14 oz.
Water................................. 4 lb.
Orange-flower water.................... 2 lb.
Alcohol............................... 3½ oz.
Oil of neroli..........................150 grains.
Tincture of vanilla .... ¾ oz.

The preceding preparations owe their activity merely to the presence of carbonate of potassium which forms an emulsion with the fat of the skin and thus resembles in its effects a mild soap. The other ingredients only serve to render the composition fragrant.

D. Toilet Powders.

Toilet powders are used to impart whiteness and smoothness to the skin; hence they are merely a kind of dry cosmetic which are applied by means of a powder puff or a hare's foot.

Their main ingredients are starch and talcum powders, perfumed and sometimes tinted a rose-red color. It is immaterial what kind of starch is used; rice, wheat, and potato starch are equally effective, provided they are clear white and in the finest powder. In some cases the bitter-almond bran remaining after the expression of the fixed oil and the preparation of the oil of bitter almond is likewise used for toilet powders. The more thoroughly these powders are rubbed into the skin, the whiter the latter becomes and the less easily can they be detected.

WHITE TOILET POWDER.

Fine levigated zinc white................	1¾ oz.
Venetian talcum......................	1¾ oz.
Carbonate of magnesia.................	1¾ oz.
Oil of rose...........................	20 drops.
Oil of orris..........................	20 drops.

Mix intimately.

PINK TOILET POWDER.

White toilet powder (see above)..........	5½ oz.
Carmine............................	8 grains.

POUDRE DE PISTACHES.

Pistachio meal.......................	10 lb.
Talcum	10 lb.
Oil of lavender......................	¾ oz.
Oil of rose..........................	½ oz.
Oil of cinnamon.....................	75 grains.

The oil must have been completely extracted from the pistachio meal, which is to be reduced to the finest powder.

POUDRE À LA ROSE.

Starch powder......................	20 lb.
Carmine...........................	¾ oz.
Oil of rose.	½ oz.
Oil of santal.......................	½ oz.
Oil of vetiver......................	150 grains.

POUDRE À LA VIOLETTE.

Starch powder .... 20 lb.
Orris root, in fine powder............. 10 lb.
Oil of bergamot...................... ¾ oz.
Ionone, 100% .... 1 oz.
Heliotropine .... 1 oz.
Oil of neroli........................ 150 grains.
Florena .... 150 grains.

POUDRE BLANCHE SURFINE (POUDRE DE RIZ).

Starch powder........................ 20 lb.
Subnitrate of bismuth................. 2 lb.
Oil of lemon......................... ¾ oz.
Oil of rose.......................... 150 grains.

BLANC DE PERLES SEC (DRY PEARL WHITE).

Venetian chalk....................... 20 lb.
Subnitrate of bismuth .... 42 oz.
Zinc white........................... 42 oz.
Oil of lemon .... 1½ oz.

ANTI-ODORIN.

Starch powder .... 1 lb.
Salicylic acid....................... 150 grains.

This mixture, which is best left unperfumed, does excellent service when used to prevent an offensive odor in stockings or shoes. The inside of the stockings is dusted with the powder, and every week a teaspoonful is sprinkled into the shoes.

SKIN GLOSS.

Carbonate of potassium.... .... 1¾ oz.
Powdered spermaceti.................. 1¾ oz.
Starch powder........................ 1 lb.
Benzoin .... ¾ oz.
Oil of bitter almond................. 150 grains.

Mix intimately and preserve in well-closed boxes. For use, stir some into water.

KALODERM.

Wheat flour	4 lb.
Almond bran	1 lb.
Orris root, in fine powder	1 lb.
Extract of rose	1 pint.
Glycerin	6 fl. oz.

Form into a dough which is thinned with water and painted on the skin.

MUSK PASTE (FOR WASHING THE HANDS).

Powdered white soap	2 lb.
Orris root, in fine powder	½ lb.
Starch powder	1½ oz.
Oil of lemon	¾ oz.
Oil of neroli	150 grains.
Tincture of musk	1½ fl. oz.
Glycerin	12 fl. oz.

Rub the starch with the glycerin in a mortar until they are thoroughly mixed. Then transfer the mixture to a porcelain capsule and apply a heat gradually raised to 284° F. (and not exceeding 290° F.), stirring constantly, until the starch granules are completely dissolved, and a translucent jelly is formed. Then gradually incorporate with it the powdered soap and orris root, and lastly the oils and tincture.

WHITENING THE HANDS.

To keep the hands soft and white, it is essential that they should not be too frequently wetted. Housework of all kinds is detrimental, and once they are thoroughly roughened it is a difficult matter to restore the natural softness and whiteness, unless a long rest is given them. However, with care and attention a good deal can be done to prevent an unsightly appearance. After any hard work they should be well cleaned with good soap and warm water, to which a little oatmeal has been added, and after being thoroughly dried the following application should be well rubbed into the skin:

WHITENING CREAM.

Zinc oxide	2 drams.
Boric acid	1 dram.
Almond oil	½ oz.
Lanolin (anhydrous)	1½ oz.
Glycerin	2 drams.
Rose water	½ oz.

Mix the lanolin and almond oil in a warm mortar, add the zinc oxide and boric acid and rub together until quite smooth; then add the glycerin and rose water. If applied at night a pair of soft kid gloves should be worn, and they should not fit too tightly.

TOILET OATMEAL.

Toilet oatmeal consists simply of fine oatmeal perfumed with some characteristic perfume, such as rose, heliotrope, violet, etc.

SKIN FOOD FOR HANDS.

Cacao butter	1 oz.
Oil of sweet almonds	1 oz.
Oxide of zinc	1 dram.
Borax	1 dram.
Boric acid	1 dram.
Oil of bergamot	6 drops.

Heat the cacao butter and oil of almonds in a double boiler, and when thoroughly blended add the zinc and borax; stir as it cools, and add the oil of bergamot last. Rub into the hands at night.

Hydrous wool fat	30 parts.
Glycerin	20 parts.
Borax	10 parts.
Eucalyptol	2 parts.
Oil of bitter almonds	1 part.

Mix thoroughly. On retiring rub the hands thoroughly and protect by wearing gloves.

CHAPTER XXXVII.

PREPARATIONS FOR THE NAILS.

NAIL SOFTENER.

To prepare the nails for manicuring, a solution of hydrogen peroxide may be used. If the nails are gently rubbed with a soft cloth dipped in the solution, and then washed with water, they may be readily manicured.

NAIL POLISHES.

French chalk is frequently employed for this purpose, or a heavy variety of precipitated chalk tinted with a little Armenian bole or carmine. It may be perfumed, if desired, with otto of roses or geranium oil. Oxide of tin diluted with twice its weight of chalk is also used as a nail polish. Equal parts of precipitated silica and prepared chalk form an excellent polisher.

FINGER-NAIL POLISHING POWDER.

Powdered tin oleate	1 oz.
Putty powder	7 oz.
Carmine	1 scruple.
Otto of rose	8 minims.
Oil of neroli	5 minims.

Triturate carefully together.

NAIL OINTMENT.

White petrolatum...................... 4 oz., av.
White Castile soap, powder .... ½ oz., av.
Oil of bergamot or other perfume........... Sufficient.

This is used for softening the nails, curing hangnails, etc.
It is to be applied at night, the fingers being covered with gloves.

LIQUID FINGER-NAIL ENAMEL.

Hard paraffin........................... 1 dram.
Oil of rose............................. 3 drops.
Chloroform............................. 2 fl. oz.

WAX POLISH FOR THE NAILS.

Eosin.................................. 10 grains.
White wax.............................. 30 grains.
Spermaceti............................. 30 grains.
White petrolatum.......................410 grains.

The important point in the manufacture of pastes of this
kind is to have the anilin dye in the finest possible state of sub-
division. It would be best, perhaps, to dissolve the dye in a
little alcohol, and incorporate the solution with the melted
petrolatum.

NAIL VARNISH.

This may be prepared by dissolving ½ dram of paraffin
wax in 2 ounces of petroleum ether. It may be tinted pink
with a little oil of alkanet, and should be applied with a camel's
hair brush. Care should be taken to keep the bottle away from
lights, as it is highly inflammable.

An inexpensive but very satisfactory nail varnish is tincture
of benzoin, applied with a camel's hair brush.

CHAPTER XXXVIII.

WATER-SOFTENERS AND BATH SALTS.

THESE consist essentially of alkaline carbonates, which act chemically on the lime salts in the water, precipitating the lime as calcium carbonate. Ordinary washing soda is quite effective in softening water, but only a small quantity is desirable for washing purposes, as an excess is apt to roughen the skin.

Sequicarbonate of soda is now largely used for softening water. It consists of small, silky needle crystals, which may be delicately perfumed, and in this form it quickly dissolves in the water. It is best kept in bottles or wooden cases, but if exposed to the air for any length of time it does not lose water and become powdery like ordinary carbonate of soda.

Borax is a milder form of water-softener, and an excess is not so harmful as in the case of carbonate of soda. It does not dissolve so readily, but in the powdered form it is perhaps more convenient than washing soda and may be sprinkled from a tin with a perforated lid into the hand basin, as required. Sometimes a mixture of powdered borax with dried carbonate of soda is used for this purpose.

Carbonate of ammonia and liquid ammonia are also employed for water-softening. The former is not so convenient to use as the liquid form, but the latter has the disadvantage of a strong smell. This is overcome in the so-called toilet ammonias by perfuming with some flower odor, such as violet.

The most suitable perfumes for water-softeners are oils of citronella, lemongrass, lavender, rosemary, and pine. A little camphor may be added if desired. Heliotropin also makes a pleasant and inexpensive perfume, only a few grains being required to a pound of sequicarbonate of soda.

EFFERVESCING BATH POWDER.

An effervescing powder for the bath may be made by the following formula:

Sodium bicarbonate......................... 85.0 parts.
Tartaric acid.............................. 71.0 parts.
Corn starch...............................113.0 parts.
Oil of lemon.............................. 0.9 part.
Oil of iris................................ 0.3 part.
Oil of cananga........................... 0.3 part.

Mix intimately. When brought in contact with water this mixture evolves carbon dioxide.

VIOLET AMMONIA

Household ammonia...................... 1 gallon.

Perfume with 4 ounces violet oil dissolved in 10 ounces alcohol, and used in a quantity to suit. Color green or purple.

CHAPTER XXXIX.

PREPARATIONS FOR THE CARE OF THE MOUTH.

BESIDES the red lips and the gums, the teeth in particular ornament the mouth. Unfortunately there are but few persons who can boast of a perfectly healthy set of teeth, which is found as a normal condition only among savages and animals. The chief causes of the admitted fact that most persons have some defect in the mouth—bad teeth, pale gums, offensive odor—lie in part in our civilization with the ingestion of hot and sometimes sour food, in part in the lack of attention bestowed on the care of the mouth by many people. The care of the mouth is most important after meals and in the morning; particles of food lodge even between the most perfect teeth and undergo rapid decomposition in the high temperature prevailing in the mouth. This gives rise to a most disagreeable odor, and the decomposition quickly extends to the teeth.

Perfectly normal healthy teeth consist of a hard, brilliant external coat, the enamel, which opposes great resistance to acid and decomposing substances. But unfortunately the enamel is very sensitive to changes of temperature and easily cracks, thus admitting to the bony part of the teeth such deleterious substances and leading to their destruction. The bulk of the tooth consists of a porous mass of bone which is easily destroyed, and thus the entire set may be lost.

Hygienic perfumery is able to offer to the public means by which a healthy set of teeth can be kept in good condition and the disease arrested in affected teeth, and by which an agreeable freshness is imparted to the gums and lips. While true perfumes may be looked upon as more or less of a luxury, the hygiene of the mouth is a necessity; for we have to deal with the health and preservation of the important masticatory apparatus which is necessary to the welfare of the whole body, so that the æsthetic factor occupies a secondary position, or rather results as a necessary consequence from a proper care of the mouth.

With no other hygienic article have so many sins been committed as with those intended for the teeth; we have had occasion to examine a number of tooth powders, some of them very high-priced, which were decidedly injurious. Thus we have known of cases in which powdered pumice stone, colored and perfumed, has been sold as a tooth powder. Pumice stone, however, resembles glass in its composition and acts on the teeth like a fine file which rapidly wears away the enamel and exposes the frail bony substance. It needs no further explanation to prove the destructive effects of such a powder on the teeth.

Many person prize finely powdered wood charcoal as a tooth powder, and to some extent they are right. Wood charcoal always contains alkalies which neutralize the injurious acids, besides traces of products of dry distillation which prevent decomposition. But these valuable properties are counteracted by the fact that charcoal is always more or less gritty, or, being insoluble, will lodge between the teeth and form the nucleus for the lodgement of other substances.

In compounding articles for the mouth and teeth—tooth powders and mouth washes—the objects aimed at are to neutralize the chemical processes that injure the teeth and gums, and to restore freshness and resisting power to the relaxed gums and mucous membranes.

Remnants of food left in the mouth after meals soon develop acids which attack the teeth; they are neutralized by basic substances or alkalies which counteract them.

The formation of organic acids from food remnants is caused by microscopic fungi (schizomycetes) which adhere to the teeth (so-called tartar) in the absence of cleanliness; against these parasites there are at our disposal a number of substances which kill them rapidly and thus for a time arrest the process of decomposition; they are therefore called antiseptics.

Another group of ingredients acts especially on such abnormal conditions of the membranous and fleshy parts of the mouth as manifest themselves by colorless, easily bleeding gums. It is mainly compounds of the tannin group which strengthen the gums and are known as astringents.

In compounding articles for the teeth it has thus far unfortunately not been customary to combine several of the substances having the above properties, the general rule being to incorporate only one in the composition, and some so-called tooth lotions consist even of aromatics alone. Such articles perfume the mouth, but have no hygienic effect upon it.

Among the essential oils, however, there is one which should form a part of every article intended for the care of the mouth, provided it can remain unchanged in the presence of the other ingredients, which would not be the case where permanagate of potassium is used. Oil of peppermint and other mint oils exert a very refreshing influence on the mucous membranes of the mouth, in which they leave a sensation of freshness lasting for some time.

We give below a number of formulas for the manufacture of articles for the care of the mouth, as to the value of which the reader can form his own opinion from what has been stated. Finally it may be observed that several of the so-called secret preparations for the care of the mouth are arrant humbugs, worthless substances being sold at exorbitant prices and, worse yet, lacking the vaunted hygienic effect owing to their chemical composition.

The articles for the care of the mouth and teeth may be divided into tooth pastes, tooth powders, tooth tinctures or lotions, and mouth washes.

A. Tooth Pastes.

TOOTH SOAP (SAVON DENTIFRICE).

Soap	2 lb.
Talcum	2 lb.
Orris root	2 lb.
Sugar	1 lb.
Water	1 lb.
Oil of clove	150 grains.
Oil of peppermint	¾ oz.

The soap should be good, well-boiled tallow soap; it is mixed with the other ingredients (the sugar is to be previously dissolved in the water) by thorough and prolonged stirring, and is usually sold in shallow porcelain boxes. The talcum or French chalk is a soft mineral with a fatty feel and is a common commercial article.

This tooth soap and other similar preparations for the care of the mouth are frequently colored rose red. Of course only harmless colors can be used. The most appropriate are rose madder lake and carmine.

TOOTH PASTE (PÂTE DENTIFRICE).

Prepared chalk..........................	2 lb.
Orris root.........	2 lb.
Sugar.........	2 lb.
Water.......	1 lb.
Madder lake............	¾ to 1½ oz.
Oil of lavender........................150 grains.
Oil of mace...........................150 grains.
Oil of clove ....150 grains.
Oil of peppermint........	1 oz.
Oil of rose...........................150 grains.

The prepared chalk used in this and many other articles is pure *precipitated* carbonate of lime. It is made from pieces of white marble, the offal from sculptors' workshops, which are placed in wide porcelain or glass vessels and covered with hydrochloric acid, when abundant vapors of carbonic acid are given off. When the development of carbonic acid has ceased, the liquid is allowed to stand at rest for several days with an excess of marble, whereby all the iron oxide is separated. This is necessary, otherwise the preparation would not be white, but yellowish. The liquid is filtered and treated with a solution of carbonate of soda (sal soda), in water as long as any white precipitate results. This precipitate is washed with pure water on a filter, and when slowly dried it forms a fine, brilliant white powder. Crystalline calcium chloride may also be purchased, dissolved in water, and treated with the soda solution to obtain the white precipitate. The quantity of madder lake in the above formula is given within the limits to form light or dark red tooth paste.

B. Tooth Powders.

QUININE TOOTH POWDER.

Prepared chalk........................	2 lb.
Starch flour..........................	1 lb
Orris root, powdered...	1 lb.
Sulphate of quinine....................	¾ oz.
Oil of peppermint.....................150 grains.

CINCHONA-BARK TOOTH POWDER.

Cinchona bark, powdered...............	1 lb.
Prepared chalk........................	2 lb.
Myrrh, powdered......................	1 lb.
Orris root, powdered..................	2 lb.
Cinnamon, powdered..................	1 lb.
Carbonate of ammonia.................	2 lb.
Oil of clove..........................	¾ oz.

BORATED TOOTH POWDER.

Borax, powered......	1 lb.
Prepared chalk........................	2 lb.
Myrrh, powdered......................	½ lb.
Orris root, powdered..................	½ lb.
Cinnamon, powdered..................	½ lb.

HOMŒOPATHIC CHALK TOOTH POWDER.

Prepared chalk........................	4 lb.
Starch flour..........................	5½ oz.
Orris root, powdered..................	½ lb.
Oil of cinnamon	1 oz.

CAMPHORATED CHALK TOOTH POWDER.

Prepared chalk........................	4 lb.
Camphor.............................	1 lb.
Orris root, powdered.......	2 lb.
Cinnamon, powdered..................	½ lb.

CHARCOAL TOOTH POWDER.

Charcoal, powdered......................	4 lb.
Cinchona bark, powered.................	1 lb.
Oil of bergamot........................	½ oz.
Oil of lemon...........................	1 oz.

The charcoal must be derived from some soft wood; willow, poplar, or buckthorn are among the most appropriate.

CUTTLEFISH-BONE TOOTH POWDER.

Prepared chalk.........................	4 lb.
Cuttlefish-bone, powdered..............	2 lb.
Orris root, powdered...................	2 lb.
Oil of bergamot.......................	¾ oz.
Oil of lemon...........................	1½ oz.
Oil of neroli...........................150 grains.
Oil of orange.........................	¾ oz.

CACHOUS AROMATISÉES.

Cachous are of a pillular composition, and used not so much for the teeth as to impart fragrance to the breath. They are made as follows:

Gum acacia............................	1½ oz.
Catechu, powdered	2¾ oz.
Licorice juice.........................	1¼ lb.
Cascarilla, powdered..................	¾ oz.
Mastic, powdered	¾ oz.
Orris root, powdered..................	¾ oz.
Oil of clove...........................	75 grains.
Oil of peppermint	½ oz.
Tincture of ambergris.................	75 grains.
Tincture of musk......................	75 grains.

Boil the solids with water until a pasty mass results which becomes firm on cooling. The aromatics are then added, and the mass is rolled into pills which are covered with genuine silver foil. One of these pills suffices to remove the odor of tobacco, etc., completely from the mouth

Pastilles Orientales.

Sugar	8 lb.
Carmine	75 grains.
Gum acacia	2 lb.
Musk	15 grains.
Oil of rose	75 grains.
Oil of vetiver	15 grains.
Civet	15 grains.
Tartaric acid	150 grains.

Add the essential oils to the powdered solids, mix intimately, and add enough water to form a stiff dough, to be made into pills which when chewed remove the odor of tobacco or other unpleasant odors.

Rose Tooth Powder.

Prepared chalk	4 lb.
Orris root, powdered	2 lb.
Madder lake	1¾ to 2½ oz.
Oil of rose	½ oz.
Oil of santal	150 grains.

Sugar Tooth Powder.

Bone-ash	4 lb.
Orris root, powdered	4 lb.
Sugar, powdered	2 lb.
Oil of bergamot	¾ oz.
Oil of citron	½ oz.
Oil of mace	75 grains.
Oil of neroli	75 grains.
Oil of orange	150 grains.
Oil of rosemary	¾ oz.

Chinese Tooth Powder.

Pumice stone	4 lb.
Starch flour	1 lb.
Madder lake	1¾ oz.
Oil of peppermint	¾ oz.

The pumice stone must be ground into the *finest* powder and levigated, before being mixed with the other ingredients. Note our remarks on pumice stone on page 258.

C. Tooth Tinctures (Lotions) and Mouth Washes (Essences Dentifrices).

EAU ANATHÉRINE.

Guaiac wood........................	3½ oz.
Myrrh	8 oz.
Cloves.............................	5½ oz.
Santal wood........................	5½ oz.
Cinnamon..........................	1¾ oz.
Alcohol...........................	4 qts.
Rose water........................	2 qts.
Oil of mace........................	75 grains.
Oil of rose........................	75 grains.
Oil of cinnamon....................	75 grains.

The solids are macerated in the alcohol, the essential oils are dissolved in the filtered liquid, and lastly the rose water is added.

EAU DE BOTOT.

This tooth tincture, which is quite a favorite, is made in different ways; the compositions made according to the French and English formulas are considered the best. For this and many other tooth tinctures rhatany root is also frequently used. Rhatany root is derived from Krameria triandra, a South American plant. Its alcoholic tincture has a red color.

A. FRENCH FORMULA.

Anise.............................	10 oz.
Cochineal..........................	¾ oz.
Mace..............................	150 grains.
Cloves............................	150 grains.
Cinnamon..........................	2¾ oz.
Alcohol...........................	3 qts.
Oil of peppermint	¾ oz.

B. English Formula.

Tincture of cedar	4 qts.
Tincture of myrrh.....................	1 qt.
Tincture of rhatany....................	1 qt.
Oil of lavender.......................	¾ oz.
Oil of peppermint	1 oz.
Oil of rose...........................	150 grains.

Borated Tooth Tincture.

Borax................................	5½ oz.
Myrrh	5½ oz.
Red santal wood......................	5½ oz.
Sugar	5½ oz.
Cologne water........................	1 qt.
Alcohol..............................	3 qts.
Water	3 pints.

Macerate the myrrh and santal wood in the alcohol, then add the Cologne water, and lastly the sugar and borax dissolved in the water.

Camphorated Cologne Water.

Camphor.............................	1 lb.
Cologne water........................	4 qts.

Cologne water with myrrh is made in the same way, by substituting a like weight of myrrh for the camphor.

Eau de Milan.

Kino	3½ oz.
Civet................................	75 grains.
Cinnamon	¾ oz.
Alcohol..............................	5 qts.
Oil of bergamot......................	150 grains.
Oil of lemon.........................	150 grains.
Oil of peppermint	¾ oz.

Kino contains an astringent, a variety of tannin, and forms a dark red solution with alcohol.

EAU DE MIALHE.

Tincture of benzoin......................	¾ oz.
Tincture of tolu........................	¾ oz.
Tincture of vanilla	150 grains.
Kino	5½ oz.
Alcohol...............................	5 qts.
Oil of anise...........................	75 grains.
Oil of peppermint	¾ oz.
Oil of star-anise.......................	75 grains.
Oil of cinnamon........................	150 grains.

MYRRH TOOTH TINCTURE.

Mace.................................	1¾ oz.
Myrrh................................	8 oz.
Cloves................................	8 oz.
Rhatany root...........................	8 oz.
Alcohol...............................	5 qts.

CHLORAL MOUTH WASH.

Chloral hydrate........................	1 oz.
Water	10 oz.

A small quantity of this, rinsed about the mouth, removes every trace of bad odor.

POTASSIUM PERMANGANATE WATER.

Potassium permanganate.................	3½ oz.
Distilled water.........................	5 qts.

Potassium permanganate easily dissolves in distilled water and forms a beautiful violet solution, a few drops of which are placed in a glass of water for use. This salt is one of the most valuable articles for the teeth; it has the property of readily giving off oxygen to organic substances and hence immediately destroys all odor in the mouth by oxidizing the organic bodies; it also removes at once the odor of tobacco smoke. After rinsing the mouth with this solution, it is well to use some peppermint water for polishing the teeth. This mouth wash leaves brown stains on linen and other materials as well as on the skin; such spots can only be removed with acids (hydrochloric, oxalic, etc.).

SALICYLATED TOOTH TINCTURE.

Salicylic acid	1¾ oz.
Orange-flower water	30 grains.
Water	2 qts.
Alcohol	1 qt.
Oil of peppermint	30 grains.

Salicylic acid is a substance possessing strong antiseptic properties; therefore, when this mouth wash is used after meals, the occurrence of any bad odor, even in persons with defective teeth, is prevented and the progress of caries is arrested, so that the acid may be considered one of the most valuable substances in hygienic perfumery.

Dissolve the salicylic acid in the warm alcohol mixed with water; add to the still warm solution the orange-flower water and the oil of peppermint dissolved in some of the alcohol.

EAU DE SALVIA.

Oil of lemon	¾ oz.
Oil of sage	1¾ oz.
Alcohol	1 qt.
Water	4 qts.

The essential oils are dissolved in the alcohol, and this solution mixed with the water.

EAU DE VIOLETTES.

Tincture of orris root	1 qt.
Rose water, triple	1 qt.
Alcohol	2 qts.
Ionone 100%	2 drams.
Heliotropin	1 oz.
Oil of bergamot	2 drams.
Oil of jasmine, syn	1 dram.

CHAPTER XL.

THE COLORS USED IN PERFUMERY.

It is now generally recognized that, next to odor, appearance plays the greatest part in the attractiveness of perfumes. In so far as appearance is dependent on bottles, packages and wrappings the subject will be touched upon elsewhere. What we are interested in at this point is the use of color in connection with perfumes.

In treating of this, while we may begin with the coloring of extracts and toilet waters, the real importance of color arises in connection with the many perfumed products which are at the same time tinted to add to their appeal to the users.

The coloring of extracts and toilet waters is a comparatively simple matter. The only requirements which must be met by the colors used are that they be soluble in the strength alcohol used, that they give a clear, pure, pleasing tint to the solutions and that they do not quickly fade into dull, unattractive shades. For this purpose vegetable colors are sometimes used with satisfaction but the modern development of the dyestuff industry has furnished numerous artificial dyes which are ideal for the perfumer's use. One of these, a particularly clear and bright green, has been known to retain its original color in perfumes more than twenty years even in cases where the perfume was exposed to direct sunlight. That this is an exceptional case will be admitted, but there are many of the artificial dyes which meet all the perfumer's requirements and give clearer tints than the vegetable dyes which were once the only ones employed. Needless to say, the dyes chosen must be those which do not stain the linen or clothing of the user.

Vegetable dyes are still used in many cases, however, particularly in preparations for the hair. This is partly because the same clearness of color tone is not demanded in these preparations and partly because the vegetable dyes used are less liable to be affected by the other ingredients present, also on account of the real or supposed advantage on the side of healthfulness.

In coloring extracts and toilet waters the most popular color seems to be a clear but not too dark green. The old rule that a perfume should be colored in accordance with the flower for which it is named no longer holds true, possibly because now so many perfumes are not named for flowers. At any rate it is not unusual to find rose or violet compositions colored green, though the more usual tints are red and violet, respectively. Lilac extracts and toilet waters are nearly all tinted with a delicate purple, which is to say a violet, while odors of the lily of the valley class are consistently colored green. Varying shades of red are frequently employed for many odors and next to green are perhaps the most attractive. One of the most modern tints and it must be admitted one of the most attractive is a clear, delicate amber.

Some consideration must be given to the container which is to be used as the more delicate shades do not appear to advantage in frosted bottles. One of the dangers to avoid, however, is that of tinting too deeply. This is particularly noticeable in cheap perfumes which seem to be trying to make up in color what they lack in odor.

The problem of coloring perfumed products is a far more complex one than that of the extracts and toilet waters. Cosmetics of all kinds, including powders and creams, are frequently though not always colored. Here the color is to be in direct contact with the skin and not only must not change in shade while mixed with the other ingredients of the cosmetic but must be absolutely non-poisonous in its nature. Most perfumers when using artificial dyes solve this latter difficulty by confining themselves largely to the so-called certified colors; that is, colors which have been passed by the United States Department of Agriculture as non-poisonous in nature and free from all poisonous ingredients. The certification is intended to permit the use of these dyes in food products but serves the useful purpose of assuring the perfumer that he need fear no ill effects from their employment in cosmetics.

The greatest care must be taken to get dyes which will not be changed by the other ingredients with which they may be in contact. Cosmetic preparations differ too widely in composition to permit any generalities as to the colors to be used. The best advice which can be given is to try out the color in the preparation in which it is intended to use it before coloring a large batch of the material.

Many though not all toilet soaps are tinted and in these aniline colors are used almost exclusively. The soap maker has few difficulties with the coloring as there are a number of dyes of tested value as soap colors which are sold as such for the purpose and he has only to choose the desired shade. A warning might be given against the danger of coloring soaps too highly.

There are so many of the artificial colors that it is impractical to attempt to give a description or even a list of those which may be used by the perfumer. Natural colors are less numerous and the following pages describe a few of the most important for the benefit of the intending user.

Yellow Colors.

Saffron.

The stigmata of Crocus sativus contain a bright yellow or orange yellow coloring matter which is easily extracted by alcohol, petroleum ether, or fat. We prefer petroleum ether in which the finely powdered saffron is macerated, the greater portion of the solvent being distilled off, and the rest of the solution is allowed to evaporate, when the pure coloring matter is left and can easily be mixed with fat. The coloring matter may also be obtained by macerating the saffron in melted lard or in olive oil.

Jonquille Pomade.

Genuine jonquille pomade, from Narcissus Jonquilla, has a handsome yellow color which is derived from the dark yellow flowers; for this reason small quantities of jonquille pomade are sometimes used for coloring pomades for the hair.

Curcuma or Turmeric.

Curcuma or turmeric root contains a very beautiful yellow coloring matter which is easily extracted by alcohol or petroleum ether. We prepare it in the same manner as stated under the head of saffron. Curcuma color cannot be used for articles containing free alkali, which changes it to brown.

RED COLORS.

Carmine.

This magnificent, though very expensive color is obtained from the cochineal insect, Coccus cacti. If good carmine is not available, a substitute may be made, for the purpose of coloring perfumery articles, by powdering cochineal, treating it with dilute caustic ammonia, and, after adding some alum solution, exposing it to the air and direct sunlight, when the coloring matter separates in handsome red flakes, which are collected and dried.

•Carthamin Red.

Safflower, the blossoms of Carthamus tinctorius, contains two coloring matters, yellow and red. The former is extracted with water from the dried flowers, and the residue is treated with a weak soda solution which dissolves the red coloring matter. When this solution is gradually diluted with acetic acid, the dye is precipitated, and after drying forms a mass with a greenish metallic lustre. This, when reduced to powder, is used for *rouge en feuilles* or *rouge en tasses*.

This coloring matter can also be prepared by introducing into the soda solution some clean white cotton on which the color is precipitated and can then be extracted with alcohol.

Alkanet.

This root, which is readily obtained in the market, contains a beautiful red coloring matter which can be extracted with petroleum ether, but is also easily soluble in fats (melted lard or warm oil). Even small amounts of it produce a handsome rose red and larger quantities a dark purple. For pomades, hair oils, and emulsions alkanet root is the best coloring matter, as it stains them rapidly, is lasting, and cheap.

Rhatany.

Rhatany root furnishes a reddish-brown coloring matter which is soluble in alcohol and is extracted with it from the comminuted root, especially for tooth tinctures and mouth washes. For the same purpose use may also be made of red santal wood and Pernambuco wood which likewise yield to alcohol, besides astringents, beautiful colors which are very suitable for such preparations.

GREEN COLORS.

Chlorophyll.

The green coloring matter of leaves is easily extracted from them, when bruised, with alcohol, and is left behind after the evaporation of the solvent. Some powders which are to have a green color are mixed directly with dried and finely divided bright green leaves such as spinach, celery, parsley leaves, etc.

For soap it is customary to use a mixture of yellow and blue which together produce a green color. Take a yellow soap, melt it, and add to its the finest powder of smalt or ultramarine until the desired tint is obtained. Indigo-carmine cannot be used, as it would impart a blue color to the skin.

BLUE COLORS.

For many preparations smalt or ultramarine is employed, but these colors are insoluble. Two soluble blue colors are aniline blue and indigo-carmine; the latter has a beautiful and intense color, but is suitable only for pomades and not for soaps because, as stated above, it would stain the skin.

Violet.

is produced by a mixture of red and blue in due proportions.

Brown.

is produced by caramel, which is made by heating sugar in an iron pot until it changes into a deep black mass which is brown only in thin threads. This color dissolves easily in water (not in alcohol) and is very suitable for soaps.

Black.

is produced by finely divided vegetable or bone black. Liquids are colored with India ink which remains suspended for a long time owing to the fine division of the carbon.

CHAPTER XLI.

MODERN TREATMENT OF NATURAL PERFUME MATERIALS.

By Dr. Eugene Charabot.

It is a delicate proposition, the extraction of perfume from the flower.

Does it not consist in capturing the soul of the flower? To appreciate the difficulties that bestrew the path leading to this goal, one should only remember the rapidity with which the flower loses its sweet and penetrating odor after it is picked. Immediately its course of life is cut short, for a process of decomposition sets in and the delicate aromatic ingredients imprisoned in the petals, etc., can not long withstand destructive influences.

The flower is something like a coquette, upon whom we have only to bring tribulation when her beauty disappears. She can not tolerate any harshness, and often the least trouble that affects her, deprives her of her charms.

However, the problem that has to be faced in luring the shy

and sweet-smelling prisoners of the petals is to science no longer a difficult one. About seventy-five years ago, in 1835 to be exact, the chemists of that day discovered that odoriferous principles of flowers dissolved out by volatile solvents and the subsequent evaporation of this solvent would leave the oil of the flower in the free state. It was necessary to find a solvent for fulfilling a number of stringent conditions, and it was not until about twenty years ago that any substantial progress was made over the first step.

The conditions to be met are the following: Perfectly dissolving the perfumes, distilling regularly within sufficiently low limits of temperature, not to exercise any chemical effect on the vegetable substance, not to leave any odor after evaporation and obtainable at a reasonable price. It was also necessary to find a device permitting of the recovery, as completely as possible, of the solvent.

The first of these problems was solved the day when the petroleum industry first furnished light bodies, perfectly rectified and deodorized. The elimination of the solvent is accomplished by distillation and can be effected completely by the use of a vacuum. Its recovery is undertaken in closed apparatus.

Devised by Robiquet, the process furnished to Massignon the first industrial results. It is well to add that as regards apparatus permitting recourse to the use and recovery of the solvent, the researches of Naudin have been attended with the happiest results.

It will suffice to consider the problem in its details, to realize the multitude of difficulties it involves, when we have in view the production of articles of irreproachable quality. The method of working, the perfection of the material, the care exercised in all the operations, the perfect knowledge required of the biochemical phenomena prevailing in the formation in the development and the modifications of the odoriferous substance, are so many factors which exercise a determinative influence in this manufacturing branch. I have also, for a great many years, devoted all my care to a study of these various conditions, which, as much in the apparatus, as even in the organic arrangement of the plant, are of a character to exercise an influence either on the perfume that is separated or on that which the growth contains.

It is to these systematic researches undertaken in the scientific laboratory and in the factory, that I attribute my success in obtaining in the method of volatile solvents the best results as regards both quality and yield as far as the state of our present knowledge is concerned.

Petroleum ether comes in contact with flowers, and is charged with their odoriferous substances. It is finally conducted into an evaporator, where it is distilled and recovered for use in another operation. As to the odoriferous products, they are obtained as residue after evaporation. They assume a more or less compact solid form, and in this condition are known as concrete essences or solid essences.

These solid essences are highly concentrated and their perfume is exactly that of the flower. If the operation of extraction has been conducted carefully and in an intelligent manner the odor has undergone no change, the perfume is a complete extract, and is not changed by any inconstant substance foreign to the products of the plant. But that it should be thus—I do not wish to repeat superfluously—it requires infinite precautions and a profound knowledge of the most favorable conditions for the treatment of each flower, as well as of the apparatus that promotes the presence of the maximum of perfume with the maximum of delicacy.

To recapitulate, the method that we are about to describe possesses the interest of furnishing products of a great delicacy, combining in small volume a considerable quantity of perfume.

If the natural essences of flowers, in their solid form, presented already some appreciable advantages over the ancient forms of products of flowers their employment gave rise to some difficulties which served to weaken some of these advantages. As a fact the petroleum ether referred to dissolves, simultaneously with the perfume of the flowers, vegetable waxes, insoluble in alcohol, absolutely non-odorous and consequently devoid of value of any kind from the point of view of a perfume. It is easy to understand that these inodorous waxes, on account of their insolubility in alcohol, make the use of the solid essences inconvenient, because in order to extract all the perfume it is necessary to undertake numerous and delicate washings.

In addition to the fact that these washings are quite a troublesome operation, they cause loss of alcohol, also loss of perfume, if the waxes are not completely removed. Finally, and this is a serious defect, they cause in the solution of odoriferous substance too high a degree of dilution. There is required really a large quantity of alcohol to remove the total volume of perfume from the inordorous wax which remains outside of the solution. We thus lose the advantage of concentration which the solid essences might offer to fall back under one of the inconveniences attendant on the use of pomades.

The method of extracting perfumes by means of volatile solvents, such as has been described, could therefore not afford to the perfumery industry the services which had been expected of it. If it may be permitted to me to mention the fact here, I have had the good fortune to be the first to remove the difficulties which have just been mentioned and to introduce in industrial practice new methods of working of a character to aid the perfumer to exercise his art with all the power of his admirable inspiration.

The period at which the process of extracting perfumes by means of volatile solvents came into the field of practical industry was about twenty years ago. The artificial perfumery industry, on the morrow of the discovery of ionone, took on a new flight, and the products of synthesis began to seek their place in the delicate compositions. Perfumery then had need of natural raw materials, sufficiently powerful and consequently sufficiently concentrated so as not to be dominated and overwhelmed by the chemical perfumes. These could introduce even into the most delicate compositions valuable properties of originality and of permanence; but on the express condition of being able to sufficiently envelop them with the aid of the product of flowers, which alone could impart delicacy and sweetness. It is this demand, further emphasized by the inclination of facing towards strong and lasting perfumes, which struck me at the same time as the troubles occasioned by the first products obtained with the aid of volatile solvents. And thus my researches were directed with a view to obtain the perfume of flowers, in the form of a powerful product, soluble in alcohol and consequently directly utilizable without previous dilution. They soon led to a satisfactory solution and to the creation of products answering to the requirements ·stated. Since then I have been able to substitute for the first processes I devised methods more perfect because derived from the total of the new knowledge gained regarding the composition of odoriferous substances as well as regarding their successive conditions in the plant.

These methods have now attained a high degree of perfection adapted to the treatment of each flower. They have enabled me, by employing suitable solvents, to leave aside, even in the course of extraction, the vegetable wax, a substance insoluble in alcohol and inodorous, and to extract solely and completely the odoriferous principle in the form of products perfectly soluble in alcohol.

These products, absolute essences of flowers (hyperessences), are exceedingly convenient to use, it being sufficient to place them in alcohol to obtain a limpid solution as concentrated as desired. Besides, the perfumer now has at his disposal elements of delicacy and sweetness sufficiently powerful to overcome even in the most delicate compositions the crudeness of artificial perfumes, which interpose like valuable ingredients of originality. By my process the absolute essence is obtained, strictly identical with the perfume of the flower and under the best possible conditions as to yield. But as exceedingly powerful products they are high-priced, according to the various flowers.

To have, corresponding with each flower, products comparable among themselves so as to make the handling and the use more convenient in creating or making up a formula it was of the greatest importance to reduce them to a uniform price. This condition is met in the liquid essences of flowers or floressences.

It is, therefore, in the form of floressences that it is—from all points of view—most advantageous and most convenient to use the odoriferous products extracted from flowers.

As I have stated above, it is a fact which is becoming more evident, that fashion tends toward strong perfumes. The perfumer must, therefore, prepare his extracts of odors in very concentrated form. Now, the products of flowers are often highly colored, and to their color is added that of the infusions of the resins, of the chemical perfumes, etc., which accompany them in the compositions. The spots which the extracts of odors usually leave on a lace handkerchief or a silk bodice are not likely to please the fair user of the perfume. Moreover, the colors which the products composing them impart to an extract are not always attractive or such as are desired, and much interest attaches to their modification, according to taste. For these reasons it was important to obtain the perfume of flowers not only in soluble form but also colorless or very slightly colored.

The problem was an exceedingly delicate one, for it was not possible to think of resorting to chemical methods, any of which would have denatured and reduced the perfume. It was necessary to avoid the formation of colorants that were not pre-existent in the flower, and to leave out the normal vegetable pigments. The solution of this problem called for prolonged research, and some ten years ago was obtained a first result in this direction, which, however, proved insufficient. More recently, with the aid of a method entirely different from that which my first attempts had followed, I have been able to solve the problem in a more satisfactory manner. Researches led to an absolutely original process, which yields the perfume of flowers *directly in uncolored form* of exceeding delicacy, surprising quality of character and absolute yield.

Consequently, the colorless essences of flowers have made rapid headway in fine perfumery, in which they have rendered valuable service.

The process has ben applied with the same success to obtaining in non-alcoholic form essences of product other than flowers, such as uncolored essence of oak moss, violet leaves, labdanum, etc., which has created for these products new and interesting openings.

It would seem that the method of volatile solvents, in the perfection of which I have endeavored to arouse interest, realizes all that can be desired. But the progress is indefinite, its limits recede in proportion as investigation tends to make them accessible. There is no problem, industrial or scientific, the solution of which does not raise a new problem. More especially my researches into the formation and the evolution of the odoriferous components of the plant led to the question whether Nature has not been too parsimonious in the distribution of her perfumes. In other words, we may ask whether it would not be possible to modify the chemical processes of life in such a manner as to make the flower more generous. This suggestion may appear somewhat Utopian, but if the idea makes way is it not true that the Utopia discovered to-day will be the reality that progress will conquer to-morrow? I have good reason for maintaining that it will be so in the present case.

We may see that in extending the domain of chemistry and of vegetable physiology the study of the subject of odoriferous substances has not only been a matter of speculation, but also of industrial application in the most immediately positive direction.

In keen competition, yet with co-operation, these two seemingly diverse industries of flower culture and synthetic chemistry have worked, ever adding new accomplishments in the domain of perfumes. On one day chemistry contributes a new note to the gamut of known perfume notes and on the next a process is perfected which enables us to capture more perfectly the ultimate delicacy of a flower essence. Thus the perfumer is ever permitted to realize new triumphs of odorant harmony, and it is easy to perceive the closeness with which the interests of these two industries are interwoven.

In fact, the development of synthetic perfumes, far from prejudicing that of the flower essences, has assisted wonderfully in their progressive evolution. For if the artificial perfumes have enabled the perfumer to attain originality and distinctiveness, the natural ones must still supply the indispensable elements of delicacy and suavity which give to the finished blend its seductive appeal.

The discovery of the synthetic perfume substances has thus broadened the field of the perfumer, and in so doing has increased enormously the demand for the natural products. The natural consequence of this has been that the methods of extraction of the flower perfumes have been developed and perfected to an extent undreamed of a generation or more ago.

The artificial perfume is powerful for the reason that it is not accompanied by any foreign substance; it is a chemical entity. It is original, for except in a few cases it is not the reproduction of any natural perfume. But for these same reasons it is brutal and lacking in finesse and delicacy, and it cannot be allowed to dominate the flower perfumes with which it may be associated.

Progress was, therefore, in a sense illusory. The essential step, the ability to extract flower perfumes in the absolute, free from all encumbranecs and inert matter, was still lacking. And it is to the solution of this difficult problem that is due the impetus which has been given to the industry of floriculture in the south of France.

Scientific problems of the natural perfume industry are not, however, confined to those of extraction. The culture and rational exploitation of the perfume-bearing plants are matters on which depend the future prosperity of the industry. It is important that the closest scrutiny be given all questions connected with the culture and collection of the raw materials.

This means that we must know more of the marvelous processes by which nature forms the odorant substances and the part they play in the life of the plant. The study embraces the formation and circulation of the constituents of the perfume bodies, the mechanism of their evolution, the genesis of the odorant materials themselves and their physiological rôle.

The distribution of the aromatic principles between the different organs of the plant gives us a distinctive means of dividing odoriferous plants into two categories. In one the essential oil makes its appearance in the green parts of the plant, and in the other it exists solely in the flowers. Thus can we consider the perfume of the entire plant and the perfume of the isolated flower.

When we take the entire plant it is observed that the odorant material appears in the young organs and continues to form and accumulate with diminishing activity until the time of flowering. Then it returns to the stem and during the period of flowering obeys the laws of diffusion. While the work of fecundation is in progress a portion of the essential oil is consumed in the inflorescences. This act accomplished, the odorant principles redescend into the stem and diffuse into the other organs, a migration provoked in part by the dessication of the inflorescences and in part by an augmentation in the osmotic pressure of the plant cells.

The practical consequence of this is that the collection of the perfume plants should be undertaken before the plants reach the consummation of the season's activity in fecundation.

When we study the production of perfumes by the flowers we find that some can continue to manufacture their odorant principles even after collection if we place them in conditions which enable them to exercise their life functions. Others of them contain at a certain period the whole of their perfume and are incapable of producing more even if their life is not arrested. The reason for this is to be found in a further study of their plant metabolism.

The esters which are such frequent and important constituents of essential oils appear to have their origin in the green parts of the plant, where they are formed from the plant acids and alcohols under the influence of the chlorophyll and a diastase which assists in the elimination of water. At the same time a portion of the alcohols becomes dehydrated, giving rise to hydrocarbons. Then during the period of fecundation the alcohols and the esters are partially oxidized, forming aldehydes and ketones and furnishing the energy which the plant requires for seed formation.

Many odorant substances, varied in their functions and diverse in their chemical structure, appear to owe their birth to the splitting up of glucosides. Even to admit of the existence of such a process enables us to formulate a satisfactory explanation for the facts observed relative to the formation of many of the essential oils and their appearance at various points in the vegetable organism.

If the glucoside encounters in the green parts of the plants conditions favorable to its breaking up the essential oil will appear immediately. But in cases where this does not occur, the glucoside in its circulation through the cells reaches the blossom, where conditions are more propitious. There the odorant principle will be liberated and only the flower will be perfumed. The phenomenon mentioned earlier of flowers which under certain conditions continue to form essence after collection and which therefore appear to live independently of the plant on which they grew is explained by a simple equilibrium in the decomposition of the glucoside.

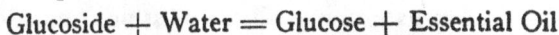

Glucoside + Water = Glucose + Essential Oil

Once an equilibrium is reached in the flower, no more essence will be formed, but let conditions be changed and the equilibrium be distributed in the right direction, and the reaction will continue until all the glucoside is broken up. So the flower, removed from its parent plant, may act as a little factory for the production of essential oil until its supply of glucoside raw material is consumed.

These researches show that, contrary to the earlier belief, the odorant principles are not useless by-products of the plant metabolism but, on the contrary, play an important rôle in its life functions, assisting in fecundation and seed growth and being partially consumed in the process.

CHAPTER XLII.

SYNTHETICS AND SPECIALTIES—THEIR PLACE IN PERFUMING.

By W. G. UNGERER.

The development of the science of structural or synthetic chemistry has placed at the disposal of the perfumer a numerous class of extremely useful and novel substances. No longer is he limited in his harmonies to the sole use of the odor notes provided by his natural perfume ingredients but by the judicious addition of the proper natural isolates and aromatic chemicals he can alter and modify his bouquets and evolve entirely new ones.

The user of these materials should employ them with discretion, however; they are valuable auxiliaries of the natural products but not substitutes for them. The perfumer who attempts to build up a fine odor solely from chemical ingredients is putting them to a use for which they were never intended and the results will be disappointing at best.

It may be said with certainty that in any fine perfume mixture floral and other natural ingredients must predominate; building on such a base the intelligent addition of aromatics will permit the attainment of wonderful effects. It is the secret of success of the great French perfumers that with a thorough understanding of the possibilities of the new ingredients they combine an appreciation of the necessity for the liberal employment of the finest floral products.

By a proper combination of the two classes of materials the perfumer of today is in a position to evolve odor harmonies distinct from any produced by nature. The consequent introduction of novel and exotic odors, many of them of extraordinary appeal, has been the dominating factor in the enormously extended use of perfumes. But the perfumer who yields to the lure of apparent economy and extends too far his employment of the achievements of the chemist at the expense of those of nature will inevitably fail of success.

But there is no greater mistake than to suppose that the chemical identities now indiscriminatingly classed as "synthetics" are all products of the laboratory built up by chemical means from simple materials. On the contrary, many of the most important are in no sense synthetics, but are merely isolated from the natural essential oils, of which they are important constituents. Thus we have citral, obtained from lemon-grass oil; geraniol, derived from citronella or geranium oil; eugenol, isolated from oil of cloves; linalool, from linaloe or bois de rose, etc.

Products of this nature demand to be placed under a separate classification, and so we are forced to divide our "synthetics" into natural isolates, or substances obtained by isolation from natural products, and aromatic chemicals, or products into whose manufacture synthetic chemistry enters.

Even in the case of these latter products the chemical process is frequently simple and consists merely in the change of a natural product into a closely related substance. Thus the natural isolate, eugenol, by a simple operation becomes the aromatic chemical, iso-eugenol, and this, in turn, is converted into vanillin. By a similarly simple conversion the citronellal of oil citronella is changed to citronellol and citral furnishes the base for ionone. Indeed, few of the so-called synthetics are not based upon isolates of natural essential oils.

While the introduction of these products has helped the perfumer, it has brought to him an entirely new set of difficult problems. The proper use of synthetics, using the term to apply to both classes, is far more difficult than the blending of nautral oils. To realize the full possibilities of a new perfume material requires frequently months and even years of experimentation by the most gifted investigators.

On account of the fact and because few perfumers can be expected to possess the time and experience required to obtain the best results the important manufacturers have found it advisable to supplement their simple products by a series of "specialties." Into the blending of these "specialties" go the artistic skill and practical experience of a gifted staff of perfumers. Several, frequently many, simple products are blended, sometimes with the admixture of natural perfumes, to give a product in which the odor note is at least partially rounded off and completed. Thus every large manufacturer of "synthetics" supplies also a line of replications of the important floral odors not necessarily complete in themselves, but furnishing valuable substitutes for the natural products with which the perfumer can build up his finished odor.

These specialties have an increasingly important place in perfuming, making available to the perfumer combinations and odor notes which he himself might never obtain, or if he did so only after arduous effort and a great expenditure of valuable time.

Some are intended to serve as substitutes for the expensive floral products, others to serve as bases for odors, while still others have their chief usefulness in rounding off the finished composition and imparting to it the almost indispensable attribute of floralcy.

For whatever purpose they are intended and used they furnish a valuable addition to the resources of the perfume, standing midway between the pure, unblended entities of the aromatic isolates and aromatic chemicals and the complex finished perfume. Few, indeed, are the perfumers who do not avail themselves to the full of the assistance offered by them.

CHAPTER XLIII.

DISCRETION IN FIXATIVES.

By W. A. PETERS.

The ideal fixative is of a retiring nature. It works behind a veil but works with power and effect. It is valued for what it does and not for what it is. It is rich in works but poor in personality. For the moment that it compels attention to its humble self it has defeated its sole purpose in existence. That purpose is to blend without intruding, to control the other ingredients in combination without masking them.

The particular quality which marks a fixative is a heaviness or strength which might be said to correspond with density in the realm of gravitation and physics. The fixative is the *deus ex machina*, the god behind the scenes, who directs the play to a happy conclusion and leaves the audience (or the *scentience*) to guess or to forget who or what it is.

The fixative is at its best when it proves capable of regulating a floral odor as well as a bouquet; when it carries its usefulness right through the life of the farewell, conserving the more volatile essences; when it rounds up the bouquet without in the least masking it.

When a perfume is in use the first element to emanate is the alcohol. After that comes the sweet flower odor, then the light and finally the heavier essential oils, all blending together into the desired grand effect, culminating in the farewell. To that bitter, or rather sweet, end the fixative must cling with tenacity to justify its place in the formula.

Discretion in the use of fixatives is the *sine qua non* of the capable perfumer. He must be discreet, not only in the type of fixative he uses, but in the quantity as well. Fixatives which serve admirably to marry heavy odors might very easily divorce the lighter, more delicate ones. And, whereas, a modicum of the selected fixative might be indispensable to the longevity of a perfume, an added modicum may kill the odor by splitting or masking it. Whereas, a perfume without a fixative is unstable and evanescent, a perfume with an overdose of the same fixative will be unbalanced, inartistic and utterly at variance with the aim of its author.

Too great heed cannot be given to the vital distinction between fixatives *per se*, which, by the way, are most desirable when most colorless, and certain essential oils which, aside from their odor value, have a more or less important fixative quality as well. In the latter case it is vital for the perfumer to take this incidental characteristic into consideration and, accordingly, modify, tone down or lessen the pure fixatives contemplated in the formula.

The animal fixatives assume first rank and importance in dealing with the binding problem in perfumes. Musk, ambergris and civet are old and tried friends. Among those of vegetable origin must be mentioned prominently tincture of orris root and the aromatic gums.

Inasmuch as it is the most essential duty of a fixative to leave unimpaired and unchanged the character of a perfume, the fatal effects of a binder subject to deterioration are manifest, for that which deteriorates in composition with other ingredients must necessarily cause deterioration, or at least change, in those .ingredients and in the combination as a whole. Thus the lasting power of a fixative is paramount to any argument which fails to take into consideration that fateful quality.

Any discussion of this sort would be incomplete without mention of the artificial musks. They enjoy a strength and timbre which make them inimitably valuable in intelligent use and vastly dangerous in the hands of the tyro. Xylene musk is known as the binder of other fixatives—indeed, it is the fixative of fixatives. Back as far as 1891, when artificial musk was in its incipient experimental stage, the late W. P. Ungerer, my preceptor in perfume art, wrote critically on this subject and discussed the need of using this aromatic chemical judiciously. His contentions have remained unshaken by the passage of years. In a spirit of common sense Mr. Ungerer tellingly made the point that to overdose a perfume with binding factors is equivalent to overdosing a human being with a medicine which in suitable proportions will cure and in greater proportions will kill.

A highly commendable, economical and widely used fixative for any perfume is orris tincture, provided that it is made from the pure Florentine article.

There are numerous well known essential oils which have too much odor character to qualify as pure fixatives but whose fixative capability is wisely recognized and employed to great advantage in formulæ to which their characteristic odors lend themselves gracefully. Employed by a discreet perfumer who understands how to temper his other fixatives to balance the heaviness of these substances, they prove both satisfactory and economical.

In casting about for fixatives bear in mind two salient points of opposite character and identical significance: Natural flower odors, by reason of their volatility have absolutely no fixative value; heavy pure fixatives like the artificial musks carry their fixative power to the point of voracity. In moderate quantity they balance the odor and add immensely to the tenacity of its farewell. In excessive quantity they eat up the delicate floral odor and produce a metallic and extremely unsatisfactory finished result.

Treat the fixative with appreciative consideration, but keep it in the background. It is necessary but not ornamental.

CHAPTER XLIV.

THE CARE AND AGING OF PERFUMES.

By L. J. ZOLLINGER.

The difficulties of the perfumer are by no means ended when he has developed a satisfactory formula and has blended choice materials with all the skill and care at his command. After this there still come the important questions of storing, aging and bottling. Many an odor has failed to develop its true excellence on account of the unreasonable haste which led to its appearance on the market too short a time after it was compounded.

But we will assume here that the reader is familiar with the advantages and indeed necessity for proper aging and discuss only the conditions under which it should be accomplished.

Even if the perfumes are to be stored in glass there are important details to be looked out for. First, it is of the utmost consequence that the bottles be carefully washed and dried in order that no particle of deleterious foreign matter be allowed to remain. Glass of suitable quality is equally essential. If the glass is of such character that it dissolves readily, for glass does dissolve in contact with liquids, particularly alcohol and water, even if the rate of solution is almost infinitely slow as compared with that of the substances which we commonly regard as soluble, it gives up some free alkali to the liquid which is almost certain to be harmful to the odor contained therein.

Another detail of consequence in this connection is the fitting of the ground glass stoppers, which must be accurate in order to avoid both loss by evaporation and damage by the entrance of air.

However, it is when we come to speak of metals in their connection with perfumes that we touch upon the subject which is most troublesome to the perfumer who has the problem of handling and storing large quantities of his finished products.

The tanking and conservation of perfumes involves a very careful study of the effects upon the product of the various metals or other materials of which tanks, pipes, funnels, funnel supports, pumps and other paraphernalia are made. Discoloration and deterioration, due to oxidation, too frequently result from ignorance of or indifference to this elemental consideration. Unsalability of a particular lot, undesirable as such an eventuation may be, is a less unfortunate happening than the loss of good will which frequently results, reacting upon the dealer and the manufacturer as well.

Of the manufacturers who actually store their perfume products in barrels—and it is a provable fact that such there are—remarks are unnecessary. We are concerned here with the less obvious and more pardonable sins of the perfumer.

As a case in point, let us take monel metal, once regarded as the alloy *par excellence* for the purposes of manufacturers of certain pharmaceutical preparations, to store essence of pepsin, for example. It was later discovered, to the chagrin of the industry generally, that monel metal contains iron, which is a dire threat to many modern perfumes containing synthetic products, such as Indol, Methyl Anthranylate, Heliotropin, Vanillin, compounds of Iso-Eugenol or any of the Salicylates.

Galvanized iron is bad, as is so-called agate ware. Enamel would be satisfactory if absolute perfection of quality were to be had in this country. Aluminum containing any trace of iron is taboo. Copper should not be employed unless heavily tin coated and, even then, it should be relined once a year, and so constructed as to permit of convenient inspection at frequent intervals. Tin cans or funnels have no status in the well appointed laboratory.

Earthenware of good quality makes excellent containers. Some perfumers here and abroad have used silver lined tanks with satisfactory results. Silver lined faucets are desirable. Brass corrodes or oxidizes and is therefore unsuitable. The top opening of the tank should be hermetically sealed against evaporation and oxidation.

Glass lined tanks are most desirable for the perfumer's every purpose. They should be thoroughly cleaned and carefully inspected before the mixing of a new batch. The slightest imperfection, even of the size of a pin-head, is liable to spell discoloration and disaster.

Rubber gaskets and washers should be employed with utmost discretion, since they dissolve in alcohol and, especially, in concentrated perfumes. Use rubber washers only where they do not come in contact with the extract, as on the outside. A ground steel V-joint is always efficient, since it can be made very tight and thus remove all danger of evaporation.

In trans-vasing, as in storing, extreme care is demanded. Never use rubber tubing, for the same reason as given in relation to rubber gaskets. Block tin or silver is appropriate to the purpose. Store the alcoholic tinctures—Musk, Civet, Ambergris —in glass demijohns away from the light and in a moderately warm temperature. Never make or store your Orris Root tincture in tin percolators and never store it in tin tanks—the extreme of folly.

Don't—if you can avoid it—allow a small quantity of a perfume to remain long in the tank. This makes for oxidation, decomposition and deterioration. It is considered good practice, for example, where the original batch is 100 gallons, to draw off and place in a suitable container the final 10 gallons.

No experienced perfumer needs to be warned to arrange his

store room of bulk perfumes in a cool, dark place, kept scrupulously clean and in apple-pie order.

The aging process depends altogether on the composition of the product. The heavy natural bouquet odors improve ten-fold by aging, whereas many new-type odors based upon artificial ingredients will not stand aging. The quicker the latter named perfumes are bottled and sold, the better. It is well for perfumers to impart this information to their clientele whenever they can do so tactfully and with the desired results.

Remember that the color of a perfumery product is all-vital—and bear in mind, in connection with the foregoing suggestions, that discoloration frequently does not take place immediately upon contact with the wrong metal but may occur long afterward, when the goods are for sale in the retail store.

A discolored product is absolutely fatal to the sale. It should be crystal-clear, well filtered, brilliant. Tint it if necessary. A light amber color is more to be desired and more salable than the dull and dirty shades which frequently emerge from the laboratory in untinted products.

When pomades were so extensively used it was recognized as vital and necessary to chill and filter the washings or flower bases. Owing to the ease in using liquid flower essences made by the volatile-solvent process, perfumers are becoming somewhat lax in this respect. But the perfumer who plays sure will test his finished product by chilling a small quantity from the batch to, say, 30°-32° F. Then and only then can he be assured that there will be no deposit in winter or in course of transportation. This precaution is not necessary in the case of all odors. It is my observation that a perfume or toilet water should stand for at least two weeks before filtering and bottling.

Simple precautions all, but redeemable in big dividends of satisfaction when you count up the losses they will save throughout a busy season in the laboratory.

CHAPTER XLV.

THE ALCOHOL PROBLEMS OF THE PERFUMER.

It is repeatedly stated elsewhere in this volume and indeed in every authoritative treatise or article on the subject that the perfumer who expects the best results in his finished extracts and toilet waters must exercise the greatest care in the selection of the alcohol used. It is repeatedly recommended that nothing but the finest cologne spirits should ever be used in a high-grade perfume. Many American perfumers would be astonished if they could witness the close attention given to this matter by the most famous and successful of the European houses. Some go so far as to say that there has never been an alcohol available in this country comparable in purity and freedom from extraneous odor with that used abroad. The writer does not feel inclined to endorse such a claim, but he is willing to go on record that far too little care is devoted to this matter here.

The perfumer who desires the best results can use nothing but the best materials, and this applies as perfectly to alcohol as to the essential oils and other materials employed. Absolute purity and absolute freedom from odor are the goals to strive for, though it is not to be expected that either will be attained easily. Insistence on finer and ever finer grades of alcohol will in time lead to an improvement if an improvement is needed as most think is the case.

Unfortunately the American perfumer encounters a special difficulty placed in his path by the Eighteenth Amendment, or rather by the manner in which it has been attempted to enforce it. Pure alcohol is an excellent substitute for the forbidden beverages when sufficiently diluted and flavored. Its inferiority to the fluids of a by-gone day is manifest, but it will serve for want of a better.

On this account and notwithstanding the fact that under the

Volstead Act it is made mandatory to encourage the extension of the use of alcohol for industrial purposes, the enforcement authorities have felt themselves compelled to restrict to the greatest possible extent the use of undenatured alcohol for any purposes whatsoever.

. The reaction on the perfume industry has been immediate. Whereas the perfumer prefers and rightly insists upon pure, undenatured alcohol he has been, in many instances, forced into accepting a denatured substitute. It cannot be too strongly urged that the industry is entitled to an adequate supply of alcohol of the quality required, but we must take circumstances as we find them sometimes, and whatever may be the opinion regarding the effect of denaturants on perfumes the average perfumer is compelled to resort to the denatured alcohol in many instances.

It is only for this reason that we discuss the use of the various denatured alcohols, and nothing which is said hereafter is to be taken as in any sense a qualification of the statement that the purest possible alcohol should be used in perfumes.

Denatured or "industrial" alcohol may be defined as pure alcohol to which certain other products have been added which render it unfit for beverage purposes. When denaturing was first permitted in this country, and indeed in other countries as well, it was to permit the use of alcohol for industrial purposes which should be free from the burdensome internal revenue taxes almost universally imposed on alcoholic beverages and on pure alcohol which, of course, is perfectly capable of easy transformation into beverages. The denaturants used at first were crude but decidedly effective, being usually possessed of a revolting odor and taste. Gasoline, bone oil and methyl or wood alcohol or combinations of these were used. Later special denaturing formulas were permitted in order that the alcohol might be used for certain purposes for which the other denaturants rendered it useless, but none of these formulas came within the exacting requirements of the perfumer.

With the advent of prohibition and the subsequent restrictions · on the use of pure alcohol attempts were made to devise formulas which would be suitable for use in perfumes, toilet waters, hair tonics, and so forth. A number of these have been officially recognized and are already in common use. In most cases they may be said to work out satisfactorily, but as regards fine perfumes it is still a subject for discussion, sometimes acrimonious, whether there is any substitute for pure alcohol.

A discussion of the various official formulas would be profitless, as they are subject to change at short notice and may be withdrawn or modified before this book is off the press. Such modifications are being made frequently in the effort to find alcohol formulas which will be as satisfactory as possible, both to the users and to the Government authorities who have ever shown a most commendable desire to co-operate in the matter. New denaturants are proposed from time to time for special purposes, and if they prove practical and satisfactory are frequently incorporated into special permitted formulas for "industrial" alcohol.

The task has been relatively simple in the case of hair tonics, mouth washes and preparations of somewhat similar character. In this class of preparations the perfume is incidental and is generally not of the sort to be particularly damaged by the addition of foreign substances to the alcohol since in many cases the perfume is used primarily to cover the inherent and not too agreeable odor of the product.

It is when we come to the finer extracts and toilet waters that the question of denaturants becomes acute. The use of the purest, undenatured alcohol is to be advocated, but when the tax on pure alcohol makes its use in all except the most expensive preparations impossible and when it is frequently difficult to get it at any price owing to permit regulations many perfumers are forced to use the denatured alcohols, made according to special formulas which are supposed to fit them for perfume use.

Two points of view, not necessarily opposed, must be met by these denaturing formulas. As the Government officials regard the question the alcohol must have added to it some ingredient or ingredients which will render it entirely unfit for use as a beverage. This does not mean that it should necessarily be violently poisonous, though several of the denatured alcohols are, but that it should possess such a nauseating or bitter flavor that no person will drink it.

This point of view must be reconciled with that of the perfumer who demands that the ingredients added be absolutely free from any odor which can conflict with that of his perfume oils and must be so chemically inert that even months of aging will not cause any unfavorable alteration in his perfected odor.

It is far too early to say that these two points of view have yet been reconciled. Specially denatured alcohols have been made available to the perfumers and have been used for some time, but there is still a division of opinion as to their success. The skilled perfumer views with no little distrust the presence of any foreign ingredient in his alcohol, however harmless it may appear. Even when it causes no immediate change in odor he fears the effects of aging.

Up to the present but one denaturing formula has met with anything like general approval. This is the formula known as number 39 B, in which the denaturant is diethyl phthalate, the diethyl ester of phthalic acid. This product, when properly prepared, is perfectly odorless, and in the percentage used, 2½ gallons to 100 gallons of alcohol, has not yet been shown to exert any deleterious effect on the finished perfumes. Judgment must be suspended, however, until trial has extended over a longer period of time and until more expert perfumers have been willing to make public their own experiences with it.

It may well be that other and better denaturants will be suggested and applied in the future. It is to be hoped, however, that such arrangements will be made eventually as will make available to the perfumer an abundant supply of pure, undenatured and tax-free alcohol.   When this is the case the troublesome problems of denaturing can be put aside and forgotten.

CHAPTER XLVI.

LEST WE FORGET.

By HENRY G. DUSENBURY.

THE two marks of a really great perfumer are Imagination and Knowledge.   The first is a heaven-sent gift; the second is largely a matter of bookkeeping.

You keep your money in the bank, not in a desk drawer. Why? Because it is precious and you want to be sure you can lay your hand on it in time of need.   Do you salt away your experiments—which represent effort, reputation, success, every-thing a perfumer has to live for—in ink-written records in a capable, business-like book?

Pencil marks are perishable and we all know what happens to scraps of paper.   The next time you hit upon a promising for-mula jot it down in lead pencil on the back of an old envelope. Then when you accidentally erase some of the data or leave the paper in one of your other suits, try to *guess* back the experi-ment.   It is an interesting game.   That way lie gray hairs in the perfume business.

Experience has taught most of us the wisdom of keeping ex-periments in an orderly book apart from the book of adopted formulas.   It doesn't pay to economize on data, including the source of supply of the ingredients used, the cost, the full name of dealer, and so on.   Some day a certain product, mellowed and improved with age, may prove the triumph of your career.   It is too big a stake to trust to memory or perishable jottings.

To refresh the memory and avoid duplication of a bad com-pound, there is nothing like an experiment book, kept neatly and in full detail. It is the best means of keeping a check on costs and results. It is the authority which enables the perfumer to say Yes or No instead of I Guess, when the head of the house wants information. It makes one sure of oneself and one's formula—which is not an undesirable state of mind for any perfumer.

Rare indeed is the perfume, soap or toilet preparation which has not in its composition some aromatic chemical or synthetic.

Hence the vital importance of examining from time to time the characteristics of artificial ingredients, with a special attention to what conditions affect them adversely in care and manipulation.

Every perfumer knows that practically all suffer from exposure to heat or light.

The experienced perfumer knows that most of them are discolored by contact with iron.

The slightest trace of rust is fatal. Hence the unwisdom of using tin cans, in which a dot of rust may hopelessly ruin a hundred gallons of a costly mixture.

In tanking or bulk-storing perfume products iron is the "chemist's itch."

The discreet perfumer uses glass, earthenware, aluminum, enamel, pure tin or extra-heavy-tinned copper for all containers and apparatus.

Experiments with monel metal have in many cases proved disastrous owing to the presence of iron in this substance.

The case comes to mind of one careful perfumer who started out with the right idea in this respect but finished in the ruck. With loving care he made a beautiful odor in glass. Much to his chagrin the product emerged a dirty brown color which defied every effort to bleach. An exhaustive investigation disclosed that he had transferred the extract via a glass funnel and a *rusty iron wire filter support!*

False economy in the buying of vessels and utensils for weighing, measuring and storing has swept away the profits of more than one odor which was inherently good and upon which a small fortune had been spent.

Uncleanly and slovenly handling has accounted for many another failure.

Don't save (?) on your apparatus. Consider the vital importance and relatively small cost of this item and get the very best that is to be had.

CHAPTER XLVII.
TOILET CREAMS.
By PETER A. FOX.

Toilet creams, like any other commodity, have a definite function to perform. This is a matter of utmost importance, yet how often is it overlooked by manufacturers when placing a cream before the public.

The several specific requirements to be covered by toilet creams are of an opposite or contradictory nature, and in consequence cannot all be properly covered by one and the same type.

The labels on many of the toilet creams on the market to-day would seem to indicate that they are a panacea, being recommended as a cleansing agent, skin food and vanishing cream, all in one. Now if we stop to consider, and our smart woman of to-day uses intelligence and does consider, the idea of using one type of toilet cream for all purposes must seem just as ridiculous to her as if she saw displayed in an exclusive boot shop a pair of riding boots with a sign attached thereto announcing: "To be worn on all occasions." Imagine her attending a function in evening attire with her feet incased in riding boots.

Milady is as critical about her complexion as she is about her attire or the food she eats. She does not consider the matter of cost, but results she must obtain. Why, then, offer her one type of cream and endeavor to convince her that she can successfully use it for all purposes? She will consider it a reflection on her intelligence and favor the manufacturer with her patronage who supplies her with the proper type of cream for each individual requirement. She will feel certain of getting positive results from this method, against negative or only doubtful results from the former. Of course, this idea can be and is carried to extremes, in which case it is even more ridiculous than the other.

One instance has come to the attention of the writer where no less than fourteen different creams were being advertised as composing a course of treatment, each having a different function to perform, and each and every one of the fourteen absolutely indispensable. We must draw our own conclusions in a case of this kind. Most likely the only differences that existed were in the perfumes with which the creams were scented.

In outlining a policy for marketing creams it seems logical to produce an individual type for each genuine requirement. The most important of these are: A cleansing cream, a skin food and a vanishing or finishing cream, and each should be prepared to do just what its name implies, nothing more or less.

Thus in preparing a cleansing cream it should be non-absorbent, oily and of a light creamy consistency, since it must be easy to manipulate it over the surface of the skin in order to collect from the pores face powder, rouge, dust or any other atmospheric impurities. If it were absorbent, instead of gathering up, it would carry minute particles of these impurities into the tissues, producing blackheads, pimples, etc.

A skin food or retiring cream must be exactly the opposite to a cleansing cream. It must be composed of rich nourishing oils and must be perfectly absorbent, otherwise it would not penetrate to the tissues, consequently could not exert the tonic, softening and nourishing effect that is required.

A vanishing cream, which may also be called a finishing or a make-up cream, because it is always employed before applying face powder, should be absolutely free from oil or grease, which would have a tendency to come through the face powder and cause the skin to become greasy or shiny, especially during warm weather. As this type is almost entirely made up from stearic acid, it is advisable to have some of it in its free (unsaponified) state, so that when it is applied to the skin, it will leave a protective film or coating on the surface which will pre vent the powder, rouge, dust and other impurities from getting directly into the pores. Furthermore, it makes a good foundation for the face powder and acts as a protection to the skin against both sun and wind.

There are other creams which enjoy a large sale, for sunburn and chap, astringent and bleach, as examples. The first is worthy of our attention, as it has a real legitimate field of usefulness. It should contain simple, mild antiseptics or medications such as boric acid, phenol, oxide of zinc, menthol, balsam of Peru and benzoin. Of the astringent and bleach creams, the less said the better. Either they do not contain the properties ascribed to them and are an imposition on the public, or, if they do contain them, they are undoubtedly harmful.

In manufacturing toilet creams only the purest of materials should be used, regardless of price. Instead of using just the ordinary white mineral oil, the type which has been purified for internal administration should be used.

The one question asked by women more often than all others combined concerning cream is: "Will it make hair or down grow on my face?" Some commonly used white mineral oils contain . sulphur. One of the most common remedies prescribed by scalp specialists for loss of hair is sulphur in the form of a lotion or an ointment. Does it not, then, seem reasonable that a cream possibly containing traces of sulphur as a heritage from the white oil from which it is made, might, in constant daily use, intensify the natural tendency of hair or down to grow upon the face? This idea may be rather far-fetched, but if there is anything in it the consequence can be avoided by using the internal oil.

Glycerine is contained in many creams, especially of the vanishing type. There are not many skins which can tolerate it. It acts as an irritant, burns and smarts, consequently not over ten per cent should be used.

The odor of a cream counts ninety per cent for or against it when offering it for sale. Let the choice of perfume be a wise one. There is nothing in the world a person will tire of more quickly than a cream distastefully perfumed. Day after day it is rubbed on the face, around the nostrils, the heat from the flesh intensifying its odor. If that odor happens to be acceptable and pleasing, it becomes an advertising asset. If, on the contrary, it contains a sickening sweet odor of the lilac or lily type so commonly encountered, it will not enjoy the steadfast patronage of the same individuals for any length of time. Consequently it will sacrifice the advertising advantages of word-of-mouth commendation, and will in many other ways lose the prestige so essential to the prolonged success of any identified article of commerce.

Let toilet creams be true creams, free from stiffness, gumminess or waxiness. While a waxy cream may successfully withstand all conditions of temperature and other factors encountered in marketing, it will, nevertheless, prove non-absorbent, and otherwise unpleasant to use, and here again the advantage of "the user becoming the advertiser" is lost.

Let the formula for a toilet cream be as simple as possible. No ingredient should be used which does not mean something to the finished product. If the scope can be fully covered by three ingredients, why use four? It simply adds to the cost of production. There have been a great many formulæ submitted to the writer for criticism, some of which have had a reputation for efficacy in producing beautiful and youthful complexions beyond the fondest dreams of Ponce de Leon. In reality these formulæ were fantasies. By eliminating four or five ingredients of no consequence the result usually simmered down to ordinary ointment of rose water, U. S. P.

The same formula would contain, for example, cucumber juice, rose water, orange flower water and distilled water. If cucumber juice, why not watermelon juice? The result is simply water. If water is desired, why go to the trouble and expense· of extracting it from cucumbers? Why, if water is the desideratum, mix this and that flower water with the water extracted from cucumbers? Why not replace them all by the cleanest, purest and simplest form of water we have, distilled water, supplying the odor represented by the this and that flower waters in the form of an attractive blend of the respective essences?

Finally the package. We are an impatient lot of humans, always more or less in a hurry, real or imaginary, and prefer things which work easily and conveniently. Our ladies much prefer a cream jar, the cover of which can be removed by simply giving, say, one-quarter turn to the left, instead of being obliged to unscrew it by giving four or five entire turns to the left. Of course, it is most important that the cover fit in such a way as to be absolutely airtight.

One of our largest manufacturing concerns several years ago in advertising one of its products alluded effectively to an improvement in the box suggesting the impossibility of improving the contents. This one thought speaks volumes. Creams should be marketed in jars from which the last particle can be conveniently removed with the fingers, not in jars that have been designed for smelling salts and require some such implement as a button hook or nail file to dig it out of the corners.

CHAPTER XLVIII.
SOAP PERFUMING.

By F. N. LANGLOIS.

There are few accomplished soap perfumers in the world, although many good perfumers. This does not indicate any essential lack on the part of the perfumer as such but merely points to the highly specialized art of perfuming soaps. The perfumer is accustomed to working with an odorless liquid solvent with a reasonable assurance that he is going to get out of it what he puts into it. This is not true of perfuming soap. One ploughs in a familiar field; the other plods through a strange jungle. The one sees what he is doing and the other works in the dark.

The perfuming of soaps involves the combining of the proposed perfume with various elements contained in the soap which are bound to react more or less upon the odor and create an effect not contemplated. It is absolutely essential that the perfumer know intimately the soap he is working upon. Of a dozen given soap stocks no two will be alike in their base, quality of vegetable oil and method of saponification, and out of the dozen no two soaps will react in the same manner to a given odor, especially where the perfume consists in part of synthetics.

Even granted that the soaps are made of the same ingredients there is always a variation in the manipulation of these ingredients and always a variation in the base. This is more especially. true since changing conditions of recent years have created new and uncertain conditions in the markets which produce high-grade fats for toilet soap making.

The importance of judicious perfuming of soap comes to mind more vividly when we reflect that very frequently the perfume content of the soap costs several times as much as the soap itself. From this we may draw the analogy of the artist who paints his masterpiece upon an inferior or faulty canvas. In the perfuming of soap two unfortunate extremes are too often present: either the soap maker, who is not necessarily a master-perfumer, being a hard-working specialist in his own line, takes upon himself the functions of perfumer with results which naturally are not always beyond criticism; or the busy expert perfumer specialist undertakes to perfume a soap with which he is not sufficiently acquainted. In either case the outcome leaves much to be desired.

In perfuming the soap the perfume manufacturer too frequently completes his odor without visualizing, as he should, a great mass of the soap. Not only should he study the completed soap compound with reference to each individual ingredient, its character, quality, proportion and stability, but he should also test each material of his perfume in combination with the soap to discover what the perfume reaction is. This is especially important in the case of synthetic compounds.

A given quantity of a particular perfume ingredient may travel a great distance in an inodorous solvent while completely losing its identity, or at least having it overwhelmingly masked, when brought into contact with the quantity of soap which the given quantity of perfume is expected to impregnate. Thus the perfumer must invariably work to the quantity and quality of the soap.

Strength is vital. The perfumer must forget that he is making a perfume and remember that he is perfuming a soap. Light floral odors will not do—they are smothered and lost. In deciding between a limited quantity of costly odor and a sufficient quantity of strength odor, the wise soap perfumer invariably chooses the latter.

The soap maker should borrow a leaf from the experience of practical perfume manufacturers—who know—and should take every precaution against overheating the product in any stage of manipulation. It is unavoidable that machinery should become somewhat heated at times, especially the mill and plodder, but the temperature should be watched and never be permitted to attain a point high enough to destroy or affect the perfume quality of the soap.

All perfume ingredients are adversely affected by heat, some more so than others. All synthetic materials, as stated above, are most susceptible to this influence. This characteristic of synthetic ingredients cannot be overemphasized. Some are merely weakened while others decompose and die. The drying room is another common danger to the perfume of the soap. Too frequently the soap is hurried into, through and out of the drying room, hastening the drying process unduly instead of allowing it to occur naturally in a moderately heated temperature.

As a final word, let me remind of the lithographer and box maker who add the bane of printers' ink to the other foes of agreeable soap perfumery. Whether, or whether not, to scent the box and wrapper rests with the individual concerned.

CHAPTER XLIX.

ALMOND OIL AND ITS SUBSTITUTES.

By W. C. ALLEN *and* E. T. BREWIS.

In speaking of almond oil we have to remember at the outset that almonds are produced in many countries, and although most of those that reach our market are shipped from a zone falling between the 30th and 45th parallels of N. latitude, which might be broadly described as "Southern Europe and countries adjacent," we have within that limit to deal with fruit produced under varying conditions both of climate and soil. This is evident when we glance at the different countries from which we get our principal supplies—viz: Morocco, Canary Islands, Portugal, Spain, France, Italy, Sicily, Syria, and Persia. Moller, in his "Lehrbuch der Pharmacognosie," states that approximately almonds contain 50 per cent. of fatty oil, but the estimate of Schaedler (45 per cent. from sweet and 38 per cent. from bitter) is more in accordance with experience.

"Almond oil" of commerce is almost entirely obtained by the expression of bitter almonds in powerful hydraulic presses. The expressed oil from "sweets" and "bitters" do not differ from each other in any material degree (compare Valencia sweets and Sicily bitters), whilst the additional product—essential oil of almonds—obtained by distillation of the "press-cake" from the latter enables the manufacturer to supply almond oil at a price that would not be possible were really "sweet almonds" alone used. We say advisedly "really" sweet almonds, because at the present time many so-called sweet almonds are being used by manufacturers which would prove very disappointing to anyone seeking a few minutes' pleasant and contemplative recreation by masticating them.

Morocco, or "Barbary," bitter almonds are always more or less mixed with sweets, and it appears to be largely a question of the paint-brush, which can readily produce an "S" or "B" according to the state of the market. The exports from the more northerly ports—viz., Saffi, Mazagan, and occasionally Rhabat—appear less open to this objection, though a slight admixture is usually met with. The supplies from Sicily are not only of larger growth, but are prepared for the market in a superior manner, being cleaner and more thoroughly sorted into their respective classes "sweet" and "bitter." Thus here, again, we must note a difference in bitter almonds, and whilst our suggested masticator would find even the ordinary Mogador "sweet" almonds bitter, if, on the other hand, he had got hold of true Sicilian bitters, his language might not bear a "qualitative analysis." The important production of the Canary Islands holds a somewhat intermediate place between that of Morocco and Sicily, whilst French, Syrian and Persian practically resemble the Sicilian almonds as regards the quality of the oil they produce. In view, therefore, of these differing sources of supply, we cannot expect absolute uniformity in results upon the examination of the various oils, especially by color-reactions; fortunately, these differences are but slight, and in no case do they reach a limit that would cause difficulty in distinguishing a genuine almond oil from one containing any of the ordinary adulterants.

The differences in the requirements of the British Pharmacopoeia and those of the United States and Germany are slight.

The American suggestion of a possibly "colorless" oil appears to foreshadow a state of perfection hardly to be anticipated here below, if we are speaking of the commercial product. The color is readily affected by prolonged exposure to light and the oil can, of course, be bleached by artificial means; concurrently with these conditions it suffers greatly in flavor.

The Brit. Pharm. 1889 states that almond oil does not congeal until nearly — 10 deg. C. The German merely says that "it remains clear at — 10 deg. C., whilst the U. S. P., with its usual thoroughness, combines the two statements. The lower limit appears to be reasonable, and conforms to our experience. All three authorities unite in giving the nitric acid test, while the German and the United States stipulate in addition a test dependent upon the melting point and solubility of the free fatty acids.

This nitric acid test has replaced that given by Bieber, who was the first, we believe, to draw attention to the means of discriminating between almond and the so-called "peach-kernel" oil. We owe much to his investigations, which brought to the knowledge of the trade how extensively the cheaper oil was being sold as "almond oil" or used to make the true "oleum amygdalae dulc." become like the product of "the widow's cruse." Although his reagent is now apt to be considered obsolete, in favor of the nitric acid test—and rightly so, we think, it is interesting to note that when comparative tests are made by the two methods, on a series of oils, some of the differing characteristics of individual samples are brought out more clearly by the Bieber test; and this is more especially noticeable after the lapse of some hours. We may, however, add that we find that the proportions 1 to 4 rather than 1 to 5 present advantages in actual working.

Maben's results, using nitric acid sp. gr. 1.42, differ from those of other observers, and have repeatedly been quoted by authors who seem to have overlooked his explanation in a later number of the journal that the oils upon which he experimented were not those usually met with in English commerce.

Later Micko pointed out that the peach-blossoms' color ascribed by Bieber to peach-kernel oil was really due to the oil from apricot-kernels. No reference to the chemical reactions of almond and kernel oils would be complete without mention of the helpful researches of Mr. J. C. Umney, who has done so much in aid of scientific production in British manufacturing pharmacy.

Since light has been thrown upon this question, we believe that adulteration of almond oil is comparatively rare. It is substitution rather than adulteration that is the practical question of the day. At the present time this question has become acute; the damage done to growing crops of almonds by the unseasonable frosts in the springs of some years has often brought about a phenomenal advance in the cost of the fruit, which has in turn affected the price of the oil. Among the possible substitutes, peach- or apricot-kernel oils stand foremost, and, indeed, are the only ones that need serious discussion. For all practical purposes oils from peach-kernels and from apricot-kernels are interchangeable. Shipments, although now consisting chiefly of apricot-kernels containing occasional packages of peach, in the past have been known to the trade as "peach-kernels," and it was this that originally guided us to use what then appeared to be the correct title for the product "oleum amygdalae persicae," from Amygdalus persica, the peach, and not, as some of our friends have freely translated it, "Persian almond oil." This oil, now so largely produced at home under its distinctive name, yet constantly described from abroad as "almond oil," has a good deal of resemblance to its illustrious namesake. It is slightly more limpid, and possesses a more nutty flavor than the true almond oil, the rich, bland, soft taste of which can be recognized by any expert.

These kernel oils have not the same keeping quality as the oil for which they are substituted, and this, together with their greater limpidity, has been the cause of various troubles where they have been unwittingly used in place of almond oil. The red color reaction of apricot, as compared with the yellowish-white of almond, is characteristic, and sharply marks a distinction between the two.

In recent years we have met with many cases of undoubted adulteration, and within the last twelve months have noticed this to a very marked extent. Out of at least seven representative sample of foreign oils obtained from different parts of the country one only could be recognized as an unsophisticated kernel oil. The principal adulterants to be looked for are oils of cotton-seed, sesame, poppy, olive and arachis. But this is a branch of the subject with which we have no practical experience, and about which we have found some difficulty in obtaining reliable information. We have, however, seen certain specimens which had nothing in common with "almond oil," whose name they bore, or of the kernel oil so often used as a substitute.

Whilst we believe that almond oil will always be facile princeps amongst fixed oils, we see no reason why the true peach- or apricot-kernel oils should not continue to find a useful place where they are suited to any particular manufacture; but we should certainly protest against any substitution of the one for the other without the knowledge of the purchaser.

CHAPTER L.

DISTINCTIVE PERFUME ADVERTISING.

By FRANCIS L. PLUMMER.

Advertising is not properly an art because it does not aim at being creative and because it is not an end in itself, but merely a means to an end. Some advertising is applied art; that is, it may, and frequently does, make incidental or collateral use of pure art for the purpose in view. But that advertising whose first aim is to command approbation for its own beauty is not sound advertising. For there is one, and only one, legitimate objective in advertising. That purpose is to attract sales, favorable notice of potential consumers, or good will to the product advertised.

Successful advertising convinces and persuades—convinces first and persuades afterward. You must make a person believe a thing before you can make him believe *in* it. You must convince him that the thing is worthy before you can persuade him to buy it. Innumerable advertising failures can be traced to overuse of the persuasive element at the expense or neglect of the convincive factor. Too much art and too little reason-why. To paraphrase—"One persuaded against his will is of the same opinion still."

In so far as impression is often fully as important as argument in the process of persuasion, beauty (art) is a factor of no mean importance in many advertising campaigns, especially in the advertising of products which themselves savor of beauty, delicacy, luxury or æstheticism.

But there comes a time or a point in every campaign where art or the impression element ceases to function advertisingly and must yield to a fuller development of the argumentative, logical or convincive element.

Advertisers frequently forget that an advertisement may be beautiful without being truly interpretive of the product advertised. They are overpersuaded by the charm of the illustration and "copy" and, perhaps subconsciously, rebuff the question: "Is this a definite as well as a beautiful message? Does it faithfully express the characteristic qualities of my product in the mood wherein it was created and in the 'manner' which is individual to its halo, its bouquet, its afterscent, its various other outstanding attributes?" They forget that beauty and art are absolute and independent, while the excellence of advertising is relative and wholly dependent upon the congeniality of the message to the product.

In the field of perfumery and related articles there is a very special and dangerous temptation to apply the art element blindly and without due consideration of its interpretive fidelity. A beauty contest ensues, rather than a merchandising race. The result is a stuffed bird masquerading as a living messenger.

Too often the character of the odor is lost in the personality of the clever men and women, generally sprayed, and not saturated, with perfume knowledge and experience, who have been delegated to spend the advertising appropriation according to their lights and undeniable talents as advertising (not perfume) specialists. Too frequently the advertiser translates too freely the *opus* of the veteran, seasoned perfumer with effects that are futile because superficial.

It is not strange that the mass of perfume advertising is more subject to this confusion of viewpoints and objectives than almost any other kind of advertising. The reason is generally recognized in what is regarded as the intangibility of this type of merchandise—an intangibility which exists solely in a state of mind and which need continue no longer than that state of mind is permitted to persist.

There must be a coming together of minds. The advertiser must learn that a perfume is a material and palpable thing and not simply an idea or effect. The perfumer must discover that advertising has passed the hokum stage and developed into an exact science with its laboratories, its micrometers and its clearly charted lanes from the source of merchandise to its destination in the consciousness of the buying public. The advertiser must become more of a perfume-artist and the perfumer must become more of an advertising scientist.

The average perfumer is at once the most material and the most volatile of men. Witness him in the laboratory as he completes the most wonderful odor of his career. He is jubilant but collected, flushed with achievement but definitely conscious of what he has done and how he has done it. He can tell you, though he won't, exactly what ingredients enter into the formula and just how many units of a particular constituent were needed to develop the grand effect. He will discourse upon the special properties of that treasured ingredient which rounded out and exquisitely finished the extract. To him there is nothing subtle or mysterious in the accomplishment. His materials and his perfected odor are very real and tangible and he can describe them to you in words that would enable you almost to sense the characteristic fragrance a block away.

Later read the advertising message of this exact scientist. Read the picture and words and seek a single hint of suggestion of the real nature of this new and characteristic fragrance. You will seek in vain. You will gather from the advertising that here is the most delicate of scents, that it has the passionate warmth of the Orient or the floral quality of sylvan dells, or that the woman who wears it is encircled with the halo of the naiads. But where, oh where! is the word or phrase to stamp this masterpiece upon the woman's cerebral processes as a definitely different, characteristic and special achievement of perfume art?

There is nothing occult or intangible about odor or the odor sense. Perfume is simply an emanation consisting of detached particles wafted from their source to the nostrils of those who happen to be in the vicinity of the source. Each particle is as real and material as the flower, bottle or box from which it is liberated. Sound exists merely in vibrations. Sight is nothing but the receptivity of the optic nerves to illusions or light. But the particles which are perfume make actual material contact with the olfactories. The sensation caused by that contact is no more mysterious than the lump raised by the impact of a brick upon the scalp of a St. Patrick's Day parader.

The reasons for the goodness or badness of an odor are similarly real and explainable. It is good if intelligently compounded from pure natural and excellent synthetic materials; better if composed from these worthy ingredients by an artist-perfumer who adds inspiration to knowledge and experience. The perfume is bad if made from indifferent or variable materials; worse if the chemist lack the skill to employ these materials in such manner as to get the poor best out of them.

Why then should it be difficult for the perfume advertiser to capitalize in his publicity the exact virtues which he knows are embraced in his product? Why not come down to earth and tell the public that it is the policy of the advertiser to choose his ingredients with tremendous care and that all natural constituents used in his merchandise are the actual odor content of the choicest flowers that Grasse can grow? Or that the particular odor is made to.the most highly prized formula ever achieved by an artist who is as famous among perfumers as Puccini is famous among musical composers? Surely women would get the drift of arguments such as these.

Just as surely as the advertising of Edison phonographs should be, and, we understand, was, built around the knowledge and theories of its illustrious inventor, just so surely should a perfume advertising campaign radiate from the man who created the odor to be advertised. He knows. He has grown up with the odor, has a very definite knowledge of why it is good, distinctive and inimitable. He is not an advertising man and cannot translate his perfume into color or form or words. But he can saturate the advertising man—the fellow who directs the writers and artists and engravers and printers—he can saturate him with his facts and moods. Then, and not before then, the advertising gentry can go to it—inspired, confident, sincere in their message to a critical public.

What are the marks by which a perfume or toilet preparation is identified, registered and retained in the buying mind? The name of the product, the name of the producer, the trademark design; all three fundamental and vital. The color and design of the bottle, box and package, a standing word, phrase or slogan; often important, but not fundamental. Variable treatments of illustration and "copy" which in time come to be associated in the popular mind with the article advertised. Here is where the efforts of the future will be more scientifically directed in the advertising of perfume products, to the end that a particular odor or preparation shall become real and tangible to the consciousness of the prospective purchaser.

The purely publicity type of advertising needs no apologies This manner of advertising bases its attack upon constant and prominent display of the trademark, name of product and of producer and has to its credit many huge successes. The weakness of purely publicity advertising is that, as competition presses, it becomes a battle of space and display. In many cases more intensive methods, along lines of educational endeavor, for example, may accomplish greater results at substantially lower cost. Individuality is the greatest of space savers.

It has been the custom to seek that saving individuality through art, aided little or not at all, by argument or reason-why appeals. But pretty faces are swiftly succeeded by prettier ones. Shapely figures and sprightly poses compete perpetually with the dryads of the Winter Garden. Art advertising is not fundamental advertising. It is merely collateral and associative, transient and too often gone without a trace behind. The bread and butter of the message must be of sterner stuff. It must be more of the convincive sort and less of the persuasive.

I think I can see signs of a new day in perfume advertising. I fancy that the advertising man is growing closer to the perfume chemist. I hope that we shall soon witness a more interesting competition of perfumes and related merchandise, which will present these products to the public in a more real and living form, not as they take shape in the fountain-pen of the "copy" specialist or on the artist's palette or in the advertiser's dream of fair women—but as they actually are present in the perfumer's formula and in the bottles upon the merchant's shelves.

That will be sincere and convincing advertising, and sincere and convincing advertising is by its very nature of the essence of good and profitable advertising.

CHAPTER LI.

CARE IN THE PRESENTATION OF PERFUMES AND TOILET PREPARATIONS.

By VICTOR VIVAUDOU.

NONE of us have the leisure or the imagination to dawdle about the average perfume display with expectant ear all cocked for the tiny plaint of, say, the Bottle of Violet, standing shyly pensive and faint in the shadow of that great bully, Tar Soap; or to loiter in the hot sun to hear the Swan Song of that Rose Odor in the window as it gently yields back its languid spirit to its heredity masters, Light and Heat.

In too many stores carrying this class of merchandise offenses pile up ad infinitum against the very simplest rules that govern the proper care and sale of the goods. To ignorance or disregard of these rules might be traced innumerable cases of sales lost and stocks damaged.

Be it known that perfumes and toilet preparations are exceedingly friendly and receptive children of nature. They take to their bosoms all manner of strange odors that are permitted to come in contact with their dainty selves. As the cuckoo deposits her brats in Mrs. Robin's nest for hatching, so does the vulgar odor slyly insinuate its particles into the composition of exposed perfume products, which assimilate the foreign smell with the avidity of a butter pat drinking the fascinating aroma of its icebox neighbor, the onion.

Light, particularly electric light, has a powerful influence over perfumes, affecting both color and odor to a marked degree. Excessive heat causes decided and rapid deterioration. It has been asserted that the rays of the moon induce fading of both perfume and label. Window displays should be carried out with these facts in mind, and it is generally well to use dummy bottles which are put up by most manufacturers for display purposes.

A too low temperature is sometimes as serious in its effects upon the appearance of perfumes as either light or heat. It pre-cipitates the solids in solution, resulting in permanent cloudiness or flakiness.

Creams perhaps suffer more severely from light, heat and cold or a sudden change of temperature than any other of the perfume and toilet preparation family. Even the very highest types of cold creams, for example, under the influence of cold, separate, granulate, dry up. Face powders are least affected of all.

It should not be a too difficult matter for the merchant to store this sort of merchandise in a temperate place and to make all other provisions necessary to protect these sensitive products from the elements. Thus he will be assured a stock that is always dainty, true to odor and to color, and will be spared many a re-turn to the manufacturer for "damages unexplained." It does no good to the joint reputation of the manufacturer and the dealer ever to sell goods that are other than fresh and attractive.

Too much care cannot be taken in handling and showing per-fumes and related articles. Isn't it illogical and wrong for the druggist to come from behind the prescription desk, unstopper a bottle of bulk perfumery and extend the odor to the customer with hands that have just been busied with iodoform or some other substance which, however grateful to cuts and sores, has nothing in common with the aesthetics of smell or with the particular scent that is being offered for sale?

And isn't it harmful neglect that allows the interchange of stoppers between two bottles, inflicting upon a Lily odor, say, the presence of a powerful Bouquet scent? This is to say nothing of the fact that each stopper is ground for its particular bottle, and, inserted in any other bottle, is almost certain to permit a leakage and evaporation of the odorous principles. Such inad-vertences as these work havoc with the sale and the goods and are not tolerated in the carefully managed store.

In the case of bulk perfumes, more commonly sold in the

West than in the East, special care is required to keep odorized fingers away from the perfume and to avoid the transposition of stoppers. This latter accident can be precluded by attaching the corks to their several bottles by, say, little aluminum chains. This should be up to the manufacturer. In handling bulk odors great care must be taken to keep the graduator clean, and when there is only a little left in the large bottle it should be decanted into a smaller container to avoid oxidation.

The samples, vials and devices supplied by progressive manufacturers should make it unnecessary to unstopper bottles on sale, and under ordinary circumstances it is well to keep the merchandise on view in the showcase rather than within the reach of prying hands.

Needless to say, much of the salability of a perfume or toilet preparation depends upon the skill and judgment with which the manufacturer has devised or selected the bottle, label and container. If he has chosen well, the task of the retailer is made infinitely more simple. But unfortunately some manufacturers fail to make the most of their opportunities in preparing the goods for market. The merchandising should proceed from the scent up and any compromise with this plan is suggestive of a friend of the writer's who has to walk stoopedly to avoid wrinkling the shoulders of a suit he bought in haste. It is not easy to produce an artistic ensemble by working backwards—and yet, they say, it is not rare for a manufacturer to select, first, the bottle, then the label, after that a box, finally, and as an incident, the *odor* to go with them!

There is nothing blatant or obtrusive about the mien of this type of merchandise, and so it is necessary to display artistry in emphasizing its presence in the window trim. The display is enhanced by having no company in the window other than draperies, jewels, works of art or such goods as suggest the dainty and aesthetic. Not too many odors should appear simultaneously but they should alternate from time to time.

The Japanese idea of decorative art has been successfully requisitioned by some Americans. In Nippon they often show in the center of the display space just one small blossom placed in a vase and set off with draperies appropriate in color and stuff to the character of the flower. A pertinent adaptation of thought might be a display comprising a single bottle of perfume or perhaps a particular series, including toilet preparations, and flanked by floral decorations—the living flower if practicable or at least a masterpiece of artificial suitable to the odor, a tawdry imitation being unthinkable for this purpose. We do not always enough appreciate the charm of simplicity.

The inside display naturally looks chiefly to location for effectiveness. Centrality, prominence, lighting advantage mean sales. But if the necessities of the business force the perfume display off into a corner, this corner should in some manner or other be enlivened to compensate.

It is difficult in many stores to set aside a special case for perfumed products, but, if it is at all possible, this space should be commandeered. For perfumes and toilet preparations orderly arrangement is Regulation Number One, and this does not exclude the tragic and unbefriended little cake of violet toilet soap that is kicked about and mutilated until it loses even its good name. Full advantage is not always taken of the artistic boxes, gotten up with great pains and at large expense for this very purpose of persuasive display. Why extract the bottle and toss these willing sales-servants into the limbo of the beneath counter region? Send them home with the bottle.

Since perfumes are so variant and elusive in their appeal, the selling of them is very largely a matter of tact and address. It is not well to force an odor upon a reluctant buyer, but it is often advisable to press the attack when the customer is inclined to reject the scent without due consideration, provided always that the sales-person is competent to demonstrate that the perfume under examination is suited to the individuality of the prospect.

In respect to the suitability of various odors to the diversities of personality, it is difficult to set down definite rules. Everyone knows that, as a general proposition, the brunette requires heavy, spicy odors and the blonde favors flowery scents, while the red-haired woman is a more difficult proposition since her personality is such as to absorb and subdue all or most types of scent. But there are other and vaguer signals of temper and manner which, though difficult to analyze, spell volumes to the initiate and make the sale easier through the application of understanding, judgment and a liberal sprinkling of intuition.

How often does a woman pooh-pooh the scent off a stopper fresh from the bottle and show delight when the demonstrator has waved away the alcohol solvent and thus discovered the odor in all its native grace! How many sales are lost by the initial proffer of a heavy odor to a woman who by all the laws of aromatics and individuality is a subject for a light floral scent! For when she afterwards arrives at the appropriate key in the series it proves insipid and futile to nostrils already cloyed with other and spicier notes of the "aromatic scale." It is a safe procedure, when in doubt, first to offer light odors with a floral halo and to conclude with a robust bouquet. And remember, the feminine viewpoint is subject to abrupt transitions.

The ideal merchant and the successful sales-person should *know* and *like* perfumes. It is not a too difficult subject to master. Any good manufacturer's salesman should be able to supply some of the basic information and tell you how and where to seek the finer details. There are but few reliable books on the subject, but ample data are available in trade papers. A little study will reflect itself in the course of selling, for nothing so thoroughly persuades a buyer as evidence of certitude on the part of the seller.

CHAPTER LII.

THE IMPORTANCE OF CORRECT PACKAGING.

By WILLIAM H. GREEN.

Standards of art and beauty vary with the habitat, race, environment and degree of education or civilization of the so-called human race. Even morals are said to be merely a matter of geography. We should not criticize the South Sea Island maid who, according to Mr. Frederic O'Brien, tattoos her legs and body for the same reason that our own civilized sister bobs her hair and shaves her eyebrows. The evident intention of both is to go nature one better and enhance the natural beauty of form or face. It is said we still have with us a few who admire, purchase and adorn the floor with the rug picturing a kind-faced St. Bernard recumbent on a center field of bilious green, which to most must be reminiscent of the sacred parlor of bygone days. It is impossible to set up a definite standard, but among the more enlightened of civilized peoples there is some general agreement as to correct line and proportion and certain colors in combination by a consensus of opinion are admitted to clash or harmonize as the case may be.

The Standard Dictionary defines Art:—"Practice as guided by correct principles in the use of means for the attainment of an end." The end to be attained by the manufacturer of perfumes and perfumed materials, hereinafter referred to as the perfumer, is naturally to adorn his products according to the dictates of the correct principles of art as understood by those to whom he expects to sell his wares.

The influence of the package in its correct relation to the product is not to be overestimated. Modern merchandisers are practically unanimous in agreement on this point. There are many illustrations with which we are all familiar. For instance, contrast the contents of the old-time weevily and musty cracker barrel with the crispy biscuit in the up-to-date, sealed, moisture-proof package. The cracker was a sufficiently commonplace article; it didn't make much headway in the barrel; but how many millions are sold now? The product only needed proper packaging to lift it from the squalid environment of the barrel in the corner grocery and put it on the shelves with the best. It has even acquired the more aristocratic title of biscuit.

The perfumer perhaps more than any other merchandiser is concerned with art and beauty in combination with practical packaging. There is a most apt illustration of the application of this in the marvelous tale of talc. A generation ago talc was hidden among the shelves of the London chemist under the title of French Chalk, occasionally sold by weight and in a paper bag, a rather plebeian article used for the same general purposes as fullers' earth and for the dry cleaning or whitening of fabrics. Adopted by the perfumer, bolted, refined, perfumed and correctly packaged it has taken a foremost place among the delicate luxuries of the toilet and probably more than one hundred and fifty million packages are sold annually in the United States alone.

Progressive package improvement, both practically and artistically, has been prominent among the factors which have advanced talc to the demand and popularity in enjoys today. Introduced in paper packages similar to those used for rice powder and just a slight improvement on the paper bag of the London chemist, talc was later and within the memory of most of us, put up in round, crudely made tins, inartistic of design and with dredge or spice box style sifter. Surely not beautiful and most inconvenient, but a step forward.

Then came the revolving sifter, a practical improvement, and year by year the tin has been improved as to shape and design, and year by year the sale of talc has increased. It is now among the aristocrats of the toilet table, packed in beautiful containers of glass, paper and tin, and sold in enormous quantities at prices commensurate with its environment. Talc needed only to be introduced to the public in a properly artistic guise to achieve its merited success.

The perfumer is continually confronted with this, to him, all important problem of correct packaging. Bottles, boxes, cartons, tins and labels make up a considerable percentage of his total purchases, the total packaging expense in many cases exceeding the cost of the contents. The importance of the expense item requires that every dollar buy its full value in package display and advertising appeal, and to do that requires that the question of art be given much intelligent thought by the perfumer himself.

Observation of the general practice of purchasing and designing containers leads to the positive conclusion that the subject is frequently given but amateur and haphazard attention. It has been a general practice to rely for the idea and its artistic realization on the manufacturer of the container. A lithographer, bottle maker, box maker or can maker is considered perforce to be an authority on art. The thought and the method are fundamentally wrong because the real incentive lies with the perfumer himself. The costume in which his product is to be introduced should be as much the child of his own thought and ideas as the product itself.

Container manufacturers as a matter of business policy have dallied with art with more or less success. The perfumer has, also, but for the most part, art as applied to containers for perfumed products has been left at the mercy of the makers of the containers. Some excellent results have been attained in this way but frequently the approach has been haphazard and blundering. We still have with us many packages that belong artistically in the period with the aforementioned St. Bernard.

Package design should be carefully studied in accordance with the generally accepted canons of art and taste. The technique of art requires of necessity years of study and intense application as well as no small degree of inborn ability and instinctive appreciation of beauty, but its principles should be familiar to the layman, particularly when they play such an important role in raising his products from mediocrity to excellence.

We need to develop systematically the interest of artists and students of art in the application of their vocation to package design. Strangely enough, no great move has been made in this direction and even those advertising agencies which have shown much interest in the application of art to the advertising of perfumes and cosmetics have paid scant attention to its potential value in improving the containers of the products they advertise.

Though by no means an exact science, even superficial study of color and design calls immediate attention to many things which surely offend good taste, that simply must not be done.

The costuming of products so intimately personal as those of the perfumer, surely demands in design the ultimate in good taste, the best thought of the man behind the product and the highest artistic efforts of those concerned in the making of the package.

Let each package be at least neutrally if not positively correct, if it be not surpassingly beautiful let it not be positively ugly. Let the perfumer and the package maker take an æsthetic as well as a practical interest in the theory of art as applied to package design. Art can be attuned to this requirement, with an aim to produce the package with the necessary advertising and display touch—art with a definite appeal, the sort of appeal that enables the prospective purchaser, momentarily forgetful of the names of both product and maker, to so accurately describe the distinctive package that the sales person is able to produce it with a smiling compliment to the purchaser on the good taste displayed in the selection.

The Golden Age of Art in Greece and Rome has handed down to us all too few of its incomparable masterpieces, but even these few furnish us with infallible examples of the correct in proportion and contour, as do the woven and mosaic arts of the

Orient in color. In the art schools and museums, everywhere for the asking is this opportunity to grasp for ourselves by observance and study and to utilize for our commercial needs a better knowledge of color and contour, so that the subtle art of the perfumer may be ably supplemented by costumes for his products, which help and enhance them and do not hinder.

Pay heed to the dictates of fashion, the current expression of the popular conception of the correct in anything. The man of civilization knows that fashion is changeable and fleeting, but the everyday things of life grow unfashionable almost unnoticed. The deservedly famous talking machine advertisement pictures a rather plump fox terrier, a dog in accordance with the supreme dictates of fashion when this brilliant idea was conceived. But today one does not notice this breed on the front seat of Milady's limousine or taking his daily airing on the avenue. He has given way to the Pekinese, the chow, the Airedale, the police dog and other canine fancies of fashion. Even the horn at which his ear is cocked has gone out of date. The advertisement would be designed differently today.

The fashion in toilet requisites is far more evanescent than in talking machines and Milady is likely to be insistent that the appointments on her boudoir table speak to the intimates who are admitted there of her familiarity with the latest edicts of Fashion. The perfumer, therefore, cannot permit that his packages fail to keep pace with the changing taste in form and color.

Many a meritorious product of the perfumer's art remains as obscure and unnoticed as a Cinderella for want of proper costuming. You will remember that even the good fairy realized the limitations imposed by the sisters' cast-off clothes and the futility of attempting to capture a Prince Charming for her protégé until Cinderella has been properly outfitted in a silken gown, glass slippers and the other necessary appendages. Then and then only was she presented for the approval of the Prince, who was evidently a discriminating person.

The moral should not be far to seek.

INDEX.

www.ingramcontent.com/pod-product-compliance
Lightning Source LLC
Chambersburg PA
CBHW020522270326
41927CB00006B/418